Performance-Based Building Design

By

Duncan Winsbury

MAPLE
PUBLISHERS

Performance-Based Building Design

Author: Duncan Winsbury

First Published in 2025

ISBN 978-1-83538-684-2 (Paperback)
978-1-83538-685-9 (Hardback)
978-1-83538-686-6 (E-Book)

Cover Design and Book Layout by:
White Magic Studios
www.whitemagicstudios.co.uk

Published by:
Maple Publishers
Fairbourne Drive, Atterbury,
Milton Keynes,
MK10 9RG, UK
www.maplepublishers.com

A CIP catalogue record for this title is available from the British Library.

Acknowledgements

I would like to thank the following for peer reviewing this book; without their sharp eyes and assistance, producing the book would have been even difficult and more time consuming.

Aidan Winsbury; *I am proud to say he is my son and a fantastic Mechanical Engineer; he works within the M&E industry in the UK. Thank you for reviewing the M&E chapter of this book, (Daniel and Rebecca, I am equally as proud of what you both have achieved in life too).*

Burhan Ghumman: *Burhan has an M.Sc. in Fire Engineering and has a bright future ahead of him; he is eager to learn and nothing is too much trouble for him. Thank you, Burhan, for reviewing the Fire Engineering chapter of this book.*

Darren Blenkinsop: *Darren is a Chartered Surveyor and a good friend; thank you for peer reviewing the Surveyors and Quantity Surveyors Chapter of this book.*

Paul Carolan: *Paul is a Project Manager with fantastic knowledge and a yen for paying attention to detail; thank you, Paul, for peer reviewing the Project Management chapter of this book.*

Saty Nar: *Saty has worked in fire enforcement for the last 15 years and is also a magistrate in the courts. Thank you, Saty, for peer reviewing the Fire Authority chapter of this book.*

Mark Hughes: *Mark is a fellow of the Institute of Workplace and Facilities Management; Mark and I have worked together on projects for many years and I regard him as a good friend. Thank you, Mark, for peer reviewing the Facilities Management Chapter of this book.*

Colleen*, my wife, for supporting me through the long hours in the production of this book, and through life as well.*

CONTENTS

Acknowledgements

I would like to thank the following for peer reviewing this book; without their sharp eyes and assistance, producing the book would have been even difficult and more time consuming.

Aidan Winsbury; *I am proud to say he is my son and a fantastic Mechanical Engineer; he works within the M&E industry in the UK. Thank you for reviewing the M&E chapter of this book, (Daniel and Rebecca, I am equally as proud of what you both have achieved in life too).*

Burhan Ghumman: *Burhan has an M.Sc. in Fire Engineering and has a bright future ahead of him; he is eager to learn and nothing is too much trouble for him. Thank you, Burhan, for reviewing the Fire Engineering chapter of this book.*

Darren Blenkinsop: *Darren is a Chartered Surveyor and a good friend; thank you for peer reviewing the Surveyors and Quantity Surveyors Chapter of this book.*

Paul Carolan: *Paul is a Project Manager with fantastic knowledge and a yen for paying attention to detail; thank you, Paul, for peer reviewing the Project Management chapter of this book.*

Saty Nar: *Saty has worked in fire enforcement for the last 15 years and is also a magistrate in the courts. Thank you, Saty, for peer reviewing the Fire Authority chapter of this book.*

Mark Hughes: *Mark is a fellow of the Institute of Workplace and Facilities Management; Mark and I have worked together on projects for many years and I regard him as a good friend. Thank you, Mark, for peer reviewing the Facilities Management Chapter of this book.*

Colleen, *my wife, for supporting me through the long hours in the production of this book, and through life as well.*

CONTENTS

Foreword

The construction and design industries are entering a new era, one shaped not by rigid adherence to prescriptive standards but by an aspiration to achieve specific, measurable outcomes across safety, sustainability, resilience, comfort, and efficiency. Performance-Based Building Design (PBBD) is at the forefront of this evolution. It is not simply a technical methodology but a philosophy of collaboration, innovation, and purpose.

This book brings together a comprehensive view of PBBD, providing not only a theoretical framework but a practical roadmap for multidisciplinary teams to design buildings that perform exceptionally throughout their lifecycle. Whether addressing the challenges of fire safety in complex environments, optimizing structural integrity under dynamic loads, or delivering energy-efficient, healthy indoor spaces, PBBD enables bespoke, high-impact solutions.

A key theme throughout this book is the interconnectivity of roles. In real-world projects, the boundaries between disciplines often blur. For example, while fire engineers lead on evacuation modelling and smoke control, architects influence these systems through spatial configuration. Similarly, mechanical engineers and sustainability consultants often collaborate on HVAC design and energy performance modelling. These overlapping responsibilities are not a complication, they are an opportunity for synergy.

However, for PBBD to succeed, one truth remains paramount: responsibilities must be clearly discussed and allocated at the earliest possible opportunity. Performance goals cannot be achieved in isolation. Early-stage planning and coordination between the client, project manager, architects, engineers, cost consultants, regulators, and end users are critical. Only by aligning expectations and assigning accountability early can teams avoid costly rework and deliver outcomes that truly meet user and societal needs.

This book is intended for everyone involved in shaping the built environment, from clients and developers to engineers, designers, contractors, surveyors, and building authorities. It invites all participants to reimagine how we define and deliver building performance, and to do so through structured collaboration, shared expertise, and bold thinking.

By embracing the principles outlined here, we have the opportunity to design not just better buildings, but better futures.

Introduction

Performance-Based Building Design (PBBD) is a progressive and holistic approach to the design and construction of buildings, rooted in the idea of achieving clearly defined performance outcomes rather than rigidly following prescriptive building codes. It empowers architects, engineers, and designers to innovate, offering flexible yet rigorous pathways to meet critical criteria such as life safety, energy efficiency, structural integrity, environmental sustainability, and long-term resilience.

Unlike traditional design approaches that focus on compliance with predefined rules, PBBD begins with the end in mind. It asks: "What should this building do under specific conditions?" rather than "What code or regulation should we follow?" This subtle shift transforms the design process from one of constraint to one of possibility, tailoring each solution to the unique challenges and aspirations of the project.

PBBD has gained prominence globally due to the increasing complexity of modern buildings, the rise in extreme environmental conditions, and growing demands for sustainable and adaptive design. It allows for innovation in materials, systems, and technologies, accommodating the evolving expectations of occupants, developers, regulators, and society at large.

Applications of Performance-Based Design

PBBD is particularly well suited to projects where conventional, prescriptive solutions are inadequate, overly restrictive, or do not fully align with the objectives of the client. It is increasingly used in the following areas:

- **High-rise and super tall buildings:** where fire safety, occupant egress, wind-induced motion, and structural performance require bespoke modelling and advanced simulation techniques.
- **Critical infrastructure and essential services:** such as hospitals, airports, military facilities, and data centres, which must continue functioning during or immediately after emergencies.
- **Complex or iconic architecture:** where form, function, and materials push the boundaries of conventional code frameworks, demanding flexible fire, structural, and environmental solutions.
- **Historic renovations and adaptive reuse:** where existing structures may not conform to modern codes, and tailored performance solutions are required to preserve heritage while enhancing safety and functionality.
- **Sustainable and net-zero buildings:** where performance metrics guide energy efficiency, material selection, and lifecycle impact in alignment with global climate goals.

Through this approach, designers can achieve a higher degree of integration between architecture, structure, services, and user experience, balancing performance, cost, and aesthetics.

The Professions Involved

Performance-based design demands a collaborative and multidisciplinary framework, where each professional brings expertise that directly contributes to achieving the performance goals. The key roles involved include:

- **Client / Developer** – Sets the vision, objectives, and funding for the project.
- **Project Managers:** Coordinate the diverse inputs, ensuring that performance benchmarks are clearly defined, regulatory pathways are navigated effectively, and timelines, budgets, and risks are managed proactively.
- **Architects:** Shape the functional and aesthetic vision of the building while ensuring that layout, materials, and geometry support performance criteria such as egress, daylighting, and thermal comfort.
- **Fire Engineers:** Design bespoke fire strategies based on modelling of fire growth, smoke movement, evacuation scenarios, and suppression system effectiveness—particularly vital for complex or high-risk structures.
- **Structural Engineers:** Use advanced analysis to ensure the building can withstand expected loads, including seismic, wind, and accidental actions, based on defined safety and serviceability goals rather than uniform code requirements.
- **Mechanical Engineer** – Designs HVAC systems to achieve thermal comfort, energy performance, and indoor air quality.
- **Electrical Engineer** – Develops lighting, power, and communication systems aligned with operational and energy goals.
- **Mechanical and Electrical (M&E) Engineers:** Develop and integrate systems for heating, ventilation, air conditioning, lighting, power, and automation that meet stringent performance targets for comfort, energy use, and carbon emissions.
- **Sustainability Consultants:** Perform simulations, modelling, and life cycle assessments to align the project with certifications such as BREEAM, LEED, or WELL, focusing on operational and embodied carbon, resource efficiency, and occupant well-being.
- **Cost Consultant / Quantity Surveyor** – Controls costs and advises on the financial impact of performance decisions.

- **Building Surveyors:** Play a vital role in inspecting, validating, and ensuring compliance with performance objectives during construction and occupancy. Their knowledge of material behaviour, building pathology, and regulation helps bridge theory and practice.
- **Facilities Manager / End User** – Provides operational insights and feedback, ensuring the design meets real-world needs.
- **Building Control** – Reviews performance-based strategies for compliance with Building Regulations and signs off on acceptability of alternatives to prescriptive codes.
- **Fire Authority** – Acts as a statutory consultee, ensuring that fire safety strategies support life safety, firefighting access, and emergency procedures.

The convergence of these disciplines under a performance-based framework encourages innovation, accountability, and evidence-based decision-making throughout the design and construction lifecycle.

A New Way Forward

Performance-Based Building Design represents not just a technical methodology but a philosophical shift in how we create the built environment. It encourages proactive problem solving, where safety, efficiency, and sustainability are achieved not by adherence to minimum standards, but through clear objectives, rational design, and measurable outcomes.

This approach is particularly relevant in today's world, where buildings must respond to increasing urbanisation, climate change, technological advancement, and the heightened expectations of users. PBBD fosters smarter, more resilient, and more adaptable buildings—buildings that perform well under stress, enhance occupant well-being., and support long-term operational goals.

This book will serve as a comprehensive guide to understanding and applying PBBD. It will explore core principles, roles and responsibilities, methodologies, and case studies, illustrating how performance-based strategies can be successfully implemented across various building types. Whether you are a designer, engineer, project manager, or stakeholder, this text aims to inform, inspire, and enable the next generation of high-performance, future-ready buildings.

Organisation Chart showing the hierarchical structure of the PBBD Team

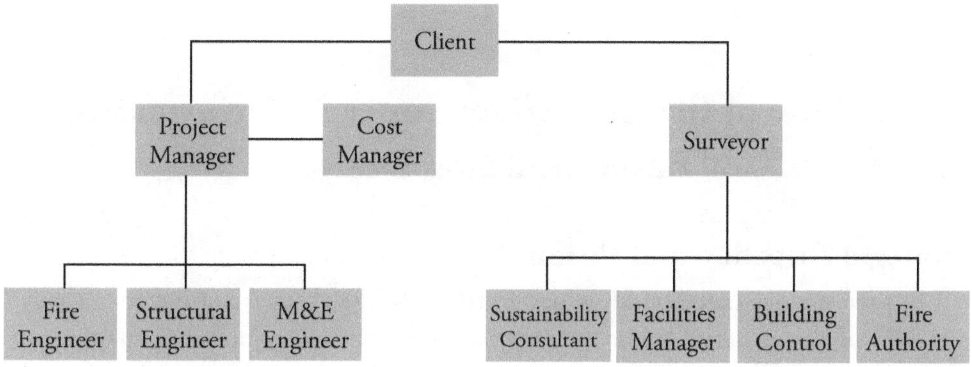

Chapter 1

The Role of the Client/Developer in Performance-Based Building Design

Strategic Direction and Vision

In performance-based building design (PBBD), the client is not merely a passive recipient of design services but an essential and active contributor to shaping the project's strategic direction. This begins with articulating clear performance aspirations that go beyond basic code compliance. Such aspirations may include achieving net-zero energy or carbon emissions, attaining enhanced fire safety standards, ensuring post-disaster operational continuity for critical infrastructure, and securing health and wellness certifications such as WELL or Fitwel.

The client's role involves translating organisational values, operational needs, and long-term business objectives into measurable performance goals. These goals form the foundation for design decisions and drive the prioritisation of systems, materials, and spatial layouts.

Beyond goal setting, forward-thinking clients act as enablers of innovation. They promote the integration of advanced materials like cross-laminated timber or carbon fibre composites, smart building technologies such as integrated sensors and AI-powered automation, and non-traditional solutions like modular construction or biophilic design. This openness to innovation encourages design teams to push boundaries and deliver customised, high-performance outcomes that are tailored to the project's context and use.

Strategically aligned clients also take into consideration broader external factors, including regulatory trends, climate change adaptation, stakeholder expectations, and the future flexibility of the building. Their vision extends across the entire lifecycle of the building, from conception and construction through to operation and maintenance, ensuring that performance-based principles are embedded at every stage.

Ultimately, a client's strategic direction and vision set the tone for the entire PBBD process, influencing procurement strategy, design choices, team collaboration, and long-term value realisation.

Early Decision-Making and Procurement Strategy

Early decision-making by the client is pivotal in laying the groundwork for a successful performance-based building design. One of the most critical early decisions is assembling a team of experts with demonstrated experience in performance-driven projects. This includes fire engineers who can conduct advanced fire modelling, structural engineers adept in non-linear and resilience-based design, mechanical and electrical (M&E) consultants who understand energy efficiency targets, and sustainability experts with BREEAM, LEED, or WELL project experience.

The client's selection process should be rooted not only in technical expertise but also in the ability to collaborate in an interdisciplinary setting. The complexity of PBBD requires seamless coordination and real-time feedback between disciplines to ensure that decisions in one domain (e.g., structural design) complement and enhance outcomes in others (e.g., fire safety or HVAC efficiency).

Procurement models are equally important. The client must select a delivery method that supports iterative design, performance modelling, and dynamic coordination. Integrated Project Delivery (IPD), early contractor involvement (ECI), or framework partnerships are preferred because they foster transparency, early engagement, and mutual accountability. These models ensure that design iterations informed by simulation outputs do not stall progress or exceed budget due to contractual rigidity.

Performance specifications should be included in all tender and consultant appointment documents. The client must clearly communicate expected performance targets, timelines for performance validation (e.g., dynamic energy modelling stages), and the criteria for acceptance or rejection. Moreover, contracts should include provisions for post-occupancy evaluation, continuous commissioning, and data transparency.

Early investment in feasibility studies, pre-construction simulations, and whole-life cost modelling gives the client a clearer view of risks and opportunities. This front-loading of analysis enables smarter decisions about systems, materials, and layout strategies, improving certainty in outcomes and cost control.

Stakeholder Engagement and Goal Alignment

The client is uniquely positioned to facilitate early alignment between project stakeholders, ensuring that performance-based goals are understood, embraced, and pursued consistently. Stakeholders may include building occupants, owners, investors, insurers, planners, and regulatory authorities—all of whom have distinct expectations and risk sensitivities.

Successful engagement involves conducting early workshops to communicate the intent behind PBBD, define shared objectives, and clarify performance expectations. For instance, insurers may be more supportive if informed about risk-reducing measures such as enhanced fire detection or redundant structural systems. Similarly, occupants can offer input on comfort expectations, which shape ventilation, lighting, and acoustic strategies.

Clients often mediate trade-offs between cost, visual design, and technical performance. For example, maintaining a building's aesthetic appeal while improving façade insulation or reducing solar gain may require iterative modelling and compromise. Clients ensure that design teams remain focused on the overarching performance goals while accommodating aesthetic, regulatory, and budgetary constraints.

Risk Tolerance and Governance

In PBBD, risk is inherent due to reliance on simulation, deviation from prescriptive codes, and the integration of innovative technologies. The client's role is to determine the organisation's risk appetite and apply it consistently across design decisions.

Clients define acceptable risk thresholds related to life safety, operational disruption, regulatory approval, and financial performance. For example, a data centre developer may prioritise redundancy and resilience over first-cost savings, leading to stricter fire protection or structural robustness targets.

Governance protocols are then established to manage these risks. Clients institute formal design gateways—milestone reviews where performance criteria, simulation results, and cost implications are assessed before proceeding. They may also create advisory panels, such as fire strategy or energy efficiency working groups, to guide decisions in complex or high-risk areas.

Performance governance extends into documentation management, peer reviews, and regulatory engagement. A clear governance structure reduces rework, ensures accountability, and supports timely decision-making.

Funding and Lifecycle Value

Clients adopting PBBD are typically focused on whole-life value rather than short-term capital savings. This perspective underpins strategic investment in systems and materials that deliver operational cost reductions, reduced insurance premiums, enhanced occupant satisfaction, and greater asset value.

They may seek ESG-linked financing, green bonds, or climate-aligned investment, which require credible documentation of energy targets, resilience measures, and sustainability outcomes. Clients must work with consultants to quantify projected savings and emissions reductions and link these to project financing models.

A lifecycle approach also informs procurement. Clients may require cost-benefit analysis for alternative HVAC systems, façade options, or lighting strategies—evaluating capital cost, maintenance, energy consumption, and system resilience. They may even commission building performance simulations for years beyond practical completion to evaluate future-proofing strategies.

Decision-Making During Design and Construction

PBBD demands an agile, informed, and involved client who can make timely decisions as new insights emerge from simulations and performance tests. The client must remain available for structured reviews and informal consultations, particularly during design coordination phases.

They must assess the trade-offs revealed by design iterations. For example, a ventilation strategy optimised for indoor air quality may increase energy consumption. The client must balance this against carbon targets and occupant well-being. priorities.

During construction, the client helps oversee change control processes. Any deviation from specified materials, systems, or layouts must be evaluated for its impact on overall building performance. The client must ensure that substitutions are validated through updated simulations and signed off by relevant consultants.

This level of involvement ensures performance goals are upheld even in the face of procurement, supply chain, or budgetary changes.

Post-Occupancy Evaluation and Continuous Commissioning

The client's responsibility doesn't end at handover. PBBD requires validation of the building's performance in use, aligning predicted and actual outcomes. Clients may commission post-occupancy evaluation (POE) studies, including occupant surveys, energy audits, IAQ testing, and system diagnostics.

They may also require Building Management Systems (BMS) to be configured for continuous monitoring of KPIs such as thermal comfort, energy consumption, CO_2 levels, and equipment uptime. Automated alerts and performance dashboards support operational tuning and fault detection.

These insights feed back into asset management strategies and future projects. Clients can create internal benchmarks or design guides based on actual building performance, institutionalising the PBBD approach.

Client in a UK Context

In the UK, clients must navigate a complex regulatory framework and integrate performance-based strategies with Building Regulations, planning requirements, and industry standards. Key areas of focus include:

- **Part L (Conservation of fuel and power):** Clients must ensure energy models comply with Part L, often using dynamic simulation modelling (DSM) tools to demonstrate energy efficiency.
- **Part B (Fire safety):** For performance-based fire strategies, clients must engage with fire engineers to produce evidence-based justification, often requiring CFD modelling, evacuation analysis, and structural fire assessments.
- **Part F (Ventilation)**: Clients must verify that ventilation strategies meet minimum requirements for IAQ while optimising for energy and comfort.

UK clients must also consider planning policy, such as local authorities' Net Zero Carbon targets or requirements for energy statements, overheating risk assessments, and BREEAM pre-assessments.

Finally, insurance and financing institutions may require third-party verification, peer reviews, and risk assessments as part of due diligence. Clients must coordinate this documentation and engage with stakeholders to secure approvals and financial support.

In summary, the UK context adds a compliance dimension to PBBD that requires rigorous technical analysis and stakeholder engagement from clients at every stage.

Chapter 2

Project Managers: The Strategic Orchestrators of Performance-Based Building Design

In performance-based building design (PBBD), the Project Manager is not simply a scheduler or budget monitor, they are the master integrator, responsible for bringing together diverse technical disciplines, aligning them with client visions, and ensuring that measurable performance objectives are achieved. Their leadership transcends traditional coordination roles to encompass foresight-driven planning, evidence-based decision-making, and continuous evaluation across all project phases.

Modern PBBD projects are often characterised by their complexity, unpredictability, and need for bespoke solutions. Within this environment, Project Managers act as both facilitators and enablers, conductors of a multi-disciplinary orchestra where every decision must harmonise with regulatory requirements, sustainability goals, and long-term operational targets. This role requires deep technical understanding, strong leadership, and an ability to balance competing priorities while maintaining alignment with performance objectives.

Defining Performance Benchmarks and Design Objectives

At the outset of a PBBD project, Project Managers take the lead in translating client expectations into actionable, measurable performance goals. This involves far more than interpreting a brief; it includes setting clear outcomes, aligning stakeholders, and building a framework through which performance can be validated. These benchmarks may include net-zero energy consumption, enhanced fire protection, thermal comfort, daylight availability, or seismic resilience.

Key responsibilities include:

- **Facilitating early stakeholder workshops** to identify strategic priorities such as climate-responsive design, health and well-being metrics, or advanced fire engineering performance.
- **Collaborating with subject matter experts** (e.g., fire engineers, sustainability consultants, M&E specialists) to convert broad ambitions into detailed Key Performance Indicators (KPIs) supported by modelling and simulation data.

- **Documenting these KPIs** in project charters, BIM execution plans, and contracts to ensure accountability and traceability throughout design and construction.
- **Revisiting performance objectives** at key design stages to accommodate new insights, changes in site conditions, or feedback from simulations.

Navigating the Regulatory Pathway

In PBBD, compliance cannot rely solely on code-checking, it often involves demonstrating that novel or non-standard solutions meet or exceed statutory requirements. Here, the Project Manager operates as a regulatory navigator, ensuring that the path to approval is strategically planned and well-justified.

Their tasks include:

- **Establishing early dialogue** with regulators, building control officers, fire authorities, and sustainability certifiers to identify potential concerns or grey areas.
- **Overseeing the preparation and submission** of robust performance documentation including computational modelling outputs, energy and fire safety simulations, evacuation analyses, and resilience reports.
- **Coordinating third-party validations** when required, such as independent fire strategy reviews or environmental impact assessments.
- **Ensuring full compliance** with UK-specific regulations such as Part L (conservation of fuel and power), Part F (ventilation), and BS 7974 (fire safety engineering), even when bespoke solutions are used.

Facilitating Interdisciplinary Collaboration

Performance-based projects require deep collaboration between disciplines. The Project Manager's role is to break down silos and establish an environment where experts can co-create without conflict or duplication. This requires a keen understanding of each discipline's needs and constraints, and the ability to synchronise their efforts.

Responsibilities include:

- **Implementing shared digital platforms**, such as Common Data Environments (CDE) or BIM coordination spaces, to promote transparency and real-time collaboration.
- **Moderating cross-functional workshops** where clashes are resolved and design opportunities are explored collectively (e.g. balancing ventilation requirements with structural fire protection).

- **Resolving design trade-offs** where objectives compete, such as prioritising thermal insulation versus fire spread risk, or daylighting versus thermal gain.
- **Maintaining collaborative momentum**, especially when teams are distributed across locations or work in parallel phases.

Managing Scope, Schedule, and Iteration

PBBD is non-linear by nature. It thrives on iteration, as performance models inform refinements and reconfiguration. The Project Manager's challenge is to introduce structure without constraining creativity, enabling fluid design evolution within firm time and cost boundaries.

Key practices include:

- **Developing dynamic project schedules** that incorporate modelling feedback loops, regulatory review periods, and staged approvals.
- **Using agile project management tools** to adaptively plan and re-forecast based on new performance data or technical setbacks.
- **Orchestrating gateway reviews** to assess readiness for progression between project stages (e.g., post-simulation sign-off before detailed design).
- **Maintaining a clear change control process** to record and justify scope evolutions driven by updated performance modelling.

Proactive Risk Management

Performance-based projects carry both conventional construction risks and additional risks unique to performance validation. Project Managers must anticipate and mitigate these risks from both technical and strategic perspectives.

Their risk leadership includes:

- **Developing and maintaining a live risk register** that specifically identifies performance risks (e.g., underperformance in thermal simulations, ventilation dead zones, structural performance gaps).
- **Quantifying risk impact** using scenario modelling and sensitivity analysis (e.g., how a shift in HVAC strategy affects carbon emissions).
- **Implementing mitigation strategies** such as back-up solutions, peer reviews, or phased testing.
- **Preparing contingency protocols** for performance failures discovered late in the design or commissioning phase.

Financial and Resource Stewardship

PBBD projects often push the limits of traditional cost planning due to their bespoke nature. Project Managers must be adept at reconciling performance ambitions with budget realities, managing both capital and lifecycle costs.

Key financial management practices include:

- **Integrating performance metrics into cost plans**, allowing design teams to explore cost-per-performance value (e.g., cost per kg of CO2 saved).
- **Leading value engineering efforts** to evaluate performance payoffs versus financial outlay across multiple options.
- **Monitoring the procurement of high-performance technologies** (e.g., adaptive glazing, high-efficiency chillers) to ensure specification compliance without overspend.
- **Optimising team resources**, ensuring that specialist consultants (e.g., façade engineers, CFD analysts) are engaged strategically, with overlapping inputs aligned to deliverables.

Quality Assurance and Performance Validation

Achieving true performance means not only designing for it but proving it. Project Managers embed validation into the project lifecycle, ensuring that the building functions as intended.

Their assurance responsibilities involve:

- **Scheduling and coordinating system commissioning activities,** including air pressure tests, sensor calibrations, and end-to-end functional testing.
- **Engaging third-party reviewers** for validation of energy models, CFD simulations, and evacuation analyses.
- **Tracking critical performance tests** (e.g., overheating assessments, fire containment trials) and comparing outcomes to initial KPIs.
- **Ensuring all evidence is documented** for final regulatory submissions and client records.

Supporting Commissioning and Post-Occupancy Evaluation

PBBD does not end with project handover. Actual performance must be monitored, calibrated, and continuously improved. The Project Manager ensures that the feedback loop remains active well into the building's operational phase.

Key post-handover activities include:

- **Coordinating soft landings and building tuning** to align systems with real-use patterns and optimise settings over time.
- **Implementing post-occupancy evaluation (POE)** protocols including occupant surveys, energy metering, and IAQ monitoring.
- **Synthesising POE data** to assess how closely the building's operation aligns with its design intent.
- **Feeding POE outcome**s back into organisational knowledge systems or future project planning.

UK Case Study: Performance-Driven Mixed-Use Development

Project: Victoria Dockside Regeneration, East London

Challenge: Design and construct a BREEAM 'Outstanding' mixed-use development with high fire safety standards, renewable energy targets, and complex occupancy patterns.

Role of Project Manager:

- Led collaborative workshops with the London Fire Brigade to integrate bespoke fire safety strategies including pressurised refuge areas and smoke control modelling.
- Coordinated multiple simulation cycles (Energy Plus and CFD) to validate energy efficiency and smoke movement.
- Managed stakeholder engagement, balancing developer ROI targets with council sustainability mandates.
- Tracked POE outcomes one-year post-completion, including EUI and occupant thermal satisfaction, feeding insights into Phase 2 of the master plan.

Performance Metrics in Project Management

Performance Category	Key Metrics Used by Project Managers
Energy Efficiency	Energy Use Intensity (EUI), kWh/m^2/year, Carbon Emission Reduction
Indoor Environment Quality	CO_2 levels (ppm), PM2.5 (μg/m^3), Operative Temperature Range
Fire Safety	Time to untenable conditions, ASET vs RSET, Evacuation time
Structural Performance	Load-resistance ratios, post-event functionality ratings
Occupant Satisfaction	Post-occupancy comfort surveys, acoustic performance
Water Efficiency	Litres/person/day, water reuse ratio

These metrics are tracked across simulations, design reviews, commissioning, and post-occupancy evaluations to ensure alignment with performance goals.

In Summary

The Project Manager in a performance-based building design project is a hybrid professional: part technical interpreter, part stakeholder diplomat, part strategist, and part guardian of outcomes. They operate at the intersection of innovation and compliance, steering projects through uncharted territory while anchoring them in sound risk management and verifiable success. By integrating design, delivery, and operation under a cohesive performance framework, they ensure that buildings are not only visionary in form but exceptional in function and endurance.

Chapter 3

The Role of the Architect in Performance-Based Building Design

Overview

In PBBD, the architect is a central figure in orchestrating spatial, aesthetic, and functional aspects of a building, while integrating complex performance criteria such as safety, sustainability, resilience, and occupant comfort. The architect becomes both a creative and technical leader, navigating between client aspirations, engineering disciplines, and regulatory frameworks.

Unlike traditional prescriptive methods, PBBD allows architects to innovate by targeting measurable performance outcomes, making the architect's early design vision critical to shaping project trajectories.

Key Responsibilities of the Architect in Performance-Based Building Design (PBBD)

Defining Performance Objectives

In performance-based building design (PBBD), the role of the architect begins not with forms or façades, but with foresight. One of the architect's most essential early-stage responsibilities is the definition and articulation of clear, quantifiable, and project-specific performance objectives. This process lays the foundation for all subsequent decisions and provides a measurable framework against which the design will be evaluated. Unlike traditional design approaches, where compliance with codes and checklists is sufficient, PBBD requires the architect to lead a process that identifies desired outcomes under actual operational conditions and sets clear targets to achieve them.

Client and Consultant Collaboration

The architect acts as a translator between client vision and technical strategy. This process typically begins during the RIBA Stage 0–1 (Strategic Definition and Preparation & Brief) and involves extensive engagement with stakeholders:

- Clients may have broad aspirations, such as achieving a net-zero carbon footprint, creating a landmark cultural facility, or developing a future-proofed healthcare environment. The architect's role is to interrogate these ambitions and distil them into technically grounded, actionable performance statements.
- User groups provide insight into operational demands, accessibility needs, health and well-being. expectations, and maintenance preferences. These perspectives are critical in shaping criteria such as lighting levels, air quality targets, and way finding strategies.
- Specialist consultants (fire engineers, M&E engineers, energy and sustainability experts, structural engineers, cost consultants) bring their domain-specific knowledge into early-stage goal setting. The architect synthesizes this input to create a holistic performance brief.

The collaborative process is iterative and requires the architect to facilitate workshops, interviews, surveys, and charrette sessions to extract relevant insights and align the vision with technical feasibility and regulatory boundaries.

Goal Setting and Prioritisation

Performance objectives in PBBD often span multiple categories, ranging from energy use and carbon reduction to resilience, safety, and occupant experience. These categories may be interrelated or even conflicting, making prioritisation essential.

The architect must manage trade-offs such as:

- Thermal comfort vs. solar gain: Maximising natural daylight can lead to overheating, which must be mitigated through façade design and ventilation strategy.
- Acoustic performance vs. ventilation: Natural ventilation strategies might conflict with acoustic insulation requirements in noisy urban areas.
- Fire resistance vs. structural innovation: Exposing structural timber for aesthetic or sustainability reasons must be balanced against its fire performance.
- To make informed decisions, architects often apply:
    - Cost-benefit analysis, considering both capital expenditure (CapEx) and operational expenditure (OpEx).
    - Life cycle thinking, evaluating the long-term implications of choices on carbon, maintenance, and adaptability.
    - Risk assessment frameworks, especially in high-occupancy or high-risk buildings, to ensure critical safety and operational functions are safeguarded.

Goals are typically ranked using a performance hierarchy:

1. Must-have outcomes (e.g., safe evacuation in under 6 minutes).
2. Targeted enhancements (e.g., BREEAM Excellent or WELL Silver).
3. Aspirational features (e.g., biophilic design, passive house standards).

Translation into Design Metrics

Once objectives are set, the architect must translate these into measurable performance indicators that guide design and enable validation. This step requires a working knowledge of modelling tools, rating systems, and performance parameters across disciplines. Examples include:

- **Thermal Comfort:** Defined through indicators like:
  - PMV (Predicted Mean Vote) and PPD (Predicted Percentage of Dissatisfied)
  - Operative temperature ranges (e.g., 21–24°C in winter; 23–26°C in summer)
  - Adaptive comfort models for naturally ventilated buildings.
- **Fire Safety:** Expressed in terms of:
  - RSET (Required Safe Egress Time) and ASET (Available Safe Egress Time)
  - Pressurisation of escape routes
  - Fire resistance ratings of critical elements (e.g., 90-minute compartmentation)
- **Energy and Emissions:**
  - Energy Use Intensity (EUI) in kWh/m^2/year
  - CO_2 emissions in kgCO_2/m^2/year (both operational and embodied)
  - Passivhaus certification standards or Net-Zero Carbon Building definitions.
- **Air Quality and Ventilation:**
  - Ventilation rates in L/s per person or m^3/h per m^2
  - CO_2 concentration thresholds (typically <1000 ppm)
  - Particulate matter (PM2.5, PM10) filtration benchmarks
- **Daylighting and Visual Comfort:**
  - Daylight Autonomy (DA), Useful Daylight Illuminance (UDI)
  - Glare Index metrics (DGP – Daylight Glare Probability)

Each of these metrics can be linked to digital models and simulations during design iterations to continuously test and refine design proposals.

Design Intent Validation

In PBBD, design is an evidence-based process. It is not enough to propose spatial arrangements or material systems that are aesthetically pleasing; they must be demonstrably capable of meeting the performance targets set in the brief.

The architect collaborates with modellers and analysts to:

- **Provide simulation-ready geometry:** This includes zoning for thermal models, escape route configurations for egress analysis, and realistic façade assemblies for daylight or energy simulations. BIM platforms often serve as the base for exporting IFC models into simulation tools such as IESVE, Energy Plus, or Fire Dynamics Simulator (FDS).
- **Coordinate feedback loops:** Modelling outputs (e.g., thermal comfort shortfalls, inadequate air changes per hour, or fire load deficiencies) prompt iterative changes to form, layout, or system design. The architect must act on this feedback while maintaining spatial and functional coherence.
- **Support peer review and third-party verification:** On complex or high-risk projects, regulators and insurers may request third-party validation of performance claims. Architects should ensure that the design documentation is transparent, well-reasoned, and traceable to performance metrics.

Ultimately, validation is not a one-time activity but a continuous dialogue between design intuition and empirical testing. Through this process, the architect becomes a steward of intent—translating aspirations into verifiable outcomes and shaping buildings that don't just look good but *perform exceptionally well.*

Integrating Multi-Disciplinary Inputs

Performance-Based Building Design (PBBD) relies on a fundamental shift from linear, discipline-siloed workflows to a dynamic, integrated model of collaboration. The architect plays a pivotal role in orchestrating this integration, ensuring that diverse technical inputs, ranging from structural resilience and fire safety to energy performance and environmental quality, are embedded within a cohesive architectural framework. This responsibility requires both technical fluency and spatial intuition, as the architect navigates the interactions between performance systems, spatial logic, and end-user experience.

Facilitating Interdisciplinary Dialogue

The architect serves as the central communication hub within the design team, enabling an ongoing exchange between a wide array of specialists, each with their own technical language, metrics, and modelling platforms. These may include:

- Structural engineers focused on load paths, member sizing, dynamic response, and robustness under extreme events.
- Fire engineers concerned with smoke control, evacuation modelling, thermal resistance, and regulatory equivalency.
- Mechanical and electrical (M&E) engineers managing ventilation, lighting, heating/cooling loads, and power distribution.
- Sustainability and environmental consultants evaluating lifecycle carbon, IAQ, water use, and user well-being.
- Acoustic consultants, lighting designers, and facade engineers addressing fine-tuned aspects of performance.
- Effective architectural leadership in this context involves:
    - Hosting and facilitating regular design coordination workshops, especially in early design stages (RIBA 1–3), where foundational strategies must be aligned.
    - Translating technical criteria into spatial opportunities, such as turning ductwork requirements into expressive ceiling elements or coordinating fire zones with intuitive way finding.
    - Mediating trade-offs between disciplines e.g., resolving conflicts between structural bracing and clear circulation paths, or between ventilation needs and acoustic attenuation in naturally ventilated classrooms.

Architects must speak the language of engineering while maintaining a design mindset, ensuring that performance requirements enhance rather than erode the spatial and experiential quality of the building.

Spatial Coordination of Systems

In PBBD, the spatial organisation of performance systems must be intentional and seamlessly integrated from the earliest design iterations. Unlike conventional prescriptive design—where systems are often overlaid late in the process, PBBD demands that space be actively shaped by performance needs, without compromising aesthetics, circulation, or usability.

Key responsibilities include:

- Mapping spatial zones for technical systems (e.g., smoke compartments, HVAC plenums, sprinkler coverage, structural bays) and ensuring their coordination with room layout, ceiling heights, service voids, and envelope articulation.
- Resolving spatial clashes between competing systems, such as duct runs crossing structural beams, or sprinkler clearance conflicting with lighting positions.
- Preserving architectural integrity while accommodating technical demands. For example, integrating structural cross-bracing into glazed atriums without compromising transparency or ensuring that smoke vents in heritage buildings are concealed within sympathetic roof details.

Through tools such as 3D coordination models, clash detection, and federated BIM environments, architects guide the spatial resolution of these systems, allowing performance engineering to support, rather than dictate, the final design form.

Interpreting Performance Data

In PBBD, the design process is data-driven and iterative. Architects are expected not only to support performance simulation efforts but to interpret and apply their results in meaningful ways, ensuring that insights gained from simulations actively inform design decisions.

Common forms of simulation and their application include:

- **CFD (Computational Fluid Dynamics) Analysis:**
    - Used to model smoke movement, ventilation efficacy, and heat transfer.
    - Informs the placement and sizing of smoke vents, pressurised stairwells, operable windows, and intake/exhaust systems.
    - Architectural implications include atrium sizing, stair pressurisation layout, and façade porosity.
- **Evacuation Modelling:**
    - Simulates occupant flow under various fire scenarios.
    - Guides the design of exit routes, stair widths, refuge areas, and horizontal evacuation strategies.
    - May prompt changes in circulation design, corridor layouts, or lift strategy in high-rise buildings.

- **Thermal and Daylight Simulations:**
  - Inform decisions about glazing ratios, façade orientation, material reflectance, shading devices, and building massing.
  - Help achieve balance between natural light, solar control, and thermal comfort, especially in offices, schools, and residential developments.
- **Structural Dynamic Analysis:**
  - Models building response to seismic, wind, or vibration loads.
  - Influences building massing, lateral load paths, bracing strategies, and damping systems.
  - May require spatial modifications to accommodate tuned mass dampers, shear cores, or base isolators.

Architects must bridge the gap between quantitative model outputs and qualitative spatial consequences, ensuring that data is not just understood but acted upon with design intelligence.

Coordinated Performance Layers

Modern buildings often require multi-functional assemblies, walls, floors, ceilings, that must simultaneously satisfy several performance domains. In PBBD, these are not viewed in isolation, but as interdependent layers whose coordination directly affects design integrity.

Examples include:

- **Façade Systems:**
  - Must deliver thermal insulation, airtightness, daylight, acoustic attenuation, and sometimes fire resistance—all while supporting architectural identity.
  - The architect coordinates cladding, glazing, shading, vapour barriers, insulation, and structural connections into a single façade strategy.
- **Floor-Ceiling Assemblies:**
  - In high-performance environments like hospitals, universities, or data centres, a single slab must accommodate:
  - Acoustic separation (e.g., impact sound insulation),
  - Fire resistance (e.g., 90-minute structural integrity),
  - Vibration control,
  - M&E service runs (e.g., ducts, cabling, sprinklers),
  - Access requirements (e.g., raised floors or access hatches).

- **Internal Partitions:**
    - Must support both acoustic and fire ratings while being reconfigurable or demountable in commercial settings.

The architect ensures that these layers are composed intentionally, often through the use of layered build-ups, systemised detailing, and sequencing strategies that allow for flexibility, inspection, and performance validation.

In essence, the architect in PBBD acts as both curator and conductor, assembling diverse technical expertise into a unified spatial and material composition that meets high-performance standards without sacrificing the coherence or creativity of architectural design.

Design for Flexibility and Innovation

Performance-Based Building Design (PBBD) fundamentally shifts the architect's role from rule-follower to performance strategist. In this framework, architects are no longer constrained by rigid compliance with prescriptive codes; instead, they are empowered to explore flexible, non-traditional, and bespoke design strategies that respond directly to a project's performance objectives. This flexibility opens the door to innovation but also places a burden of rigorous justification on every design decision.

In PBBD, innovation is not about novelty for its own sake; it is about optimising building performance, whether in energy use, occupant safety, climate adaptability, or long-term resilience, through creative, technically grounded solutions.

Non-Prescriptive Strategies

PBBD liberates architects from 'tick-box' design and instead invites them to create customised pathways to meet functional and regulatory goals. The architect must work from first principles, starting with performance intent and building up the solution accordingly.

Examples of non-prescriptive strategies include:

- **Passive Solar and Environmental Design:**
    - Buildings are oriented and massed to harness natural daylight and solar gain in winter, while using deep overhangs, louvres, or brise-soleils to prevent summer overheating.
    - Façades are dynamically tuned to context: highly glazed on the south with shading, more insulated and solid on the north.
    - Instead of simply meeting U-values or ventilation rates, the architect designs with climate responsiveness as a core strategy.

- **Adaptive Reuse and Retrofit Innovation:**
  - Existing buildings are retrofitted to meet modern performance criteria through smart layering of insulation, façade upgrades, compartmentation strategies, and active system overlays.
  - Fire strategies for heritage buildings may include pressurised escape paths or aspirating smoke detection in concealed voids.Instead of replacing structures, the architect extends the lifespan of existing fabric, contributing to sustainability and heritage preservation.
- **Advanced Egress Concepts:**
  - Particularly in high-rise, large-assembly, or mixed-use buildings, traditional escape stairs may be inadequate or spatially unfeasible.
  - Architects can propose phased evacuation, refuge floors, pressurised lobbies, or internal fire-rated escape tunnels, all supported by modelling and risk assessment rather than prescriptive stair count requirements.
  - The focus shifts from static code adherence to dynamic evacuation performance under real-world scenarios.

Such strategies must be tested through simulation and approved through performance justification, shifting the architect's design process from "comply and proceed" to "design, test, and prove."

Smart and Responsive Systems

Innovation in PBBD often involves the integration of intelligent building systems that adapt in real time to changing conditions, enhancing efficiency and occupant well-being.

- **Dynamic Glazing:** Electrochromic or thermochromic glass adjusts transparency based on sunlight, balancing glare reduction and thermal comfort without blinds or mechanical shading.
- **Building-Integrated Photovoltaics (BIPV):** PV elements are integrated directly into the envelope, façades, skylights, or canopies, contributing to both energy generation and aesthetic identity.
- **Automated Shading and Climate Controls:**
  - Shading systems, lighting, and HVAC are interconnected through Building Management Systems (BMS) that respond to sensor data on occupancy, daylight levels, and temperature.

- Zoning strategies may be based on user behaviour and space usage patterns, with demand-controlled ventilation and lighting optimising performance at the micro level.

- **Real-Time Monitoring and Feedback Loops:**
  - Systems provide live data on indoor air quality, energy use, and environmental comfort, enabling building operators to fine-tune systems, and architects to learn from post-occupancy behaviour for future designs.

By designing for system intelligence and responsiveness, architects transform buildings into active participants in their own performance.

Engineering Judgement and Scenario Planning

A hallmark of PBBD is its use of performance scenarios to stress-test the design against a range of conditions, from routine operations to extreme or unforeseen events.

The architect works closely with fire engineers, structural consultants, and risk specialists to:

- Develop multiple use-case scenarios, including:
  - Best-case: favourable conditions with expected occupancy.
  - Typical case: standard operational mode.
  - Worst-case: peak occupancy with equipment failure, fire ignition, or environmental extremes.
- Design for system robustness, incorporating:
  - Fire redundancy: secondary egress routes, dual water supplies for sprinklers, or intelligent smoke detection to account for false negatives.
  - Structural continuity: designing for progressive collapse prevention, often via load redistribution or sacrificial elements.
  - Resilient MEP zoning: electrical and ventilation systems designed in zones so that partial outages do not compromise total building function.

These scenarios are then modelled and analysed using tools like CFD (smoke movement), dynamic structural simulation, and evacuation modelling. The architect must interpret these results, adjusting spatial layout, material choices, or system design to mitigate risk and deliver performance across the lifecycle, not just under ideal conditions.

Material Innovation

Materials in PBBD are selected not just for appearance, cost, or availability, but also for their performance potential, lifecycle impact, and alignment with sustainability or resilience goals. The architect explores next-generation materials through both creative design and scientific reasoning.

Examples include:

- Cross-Laminated Timber (CLT):
  - Offers carbon sequestration, low embodied energy, and fast assembly.
  - Fire performance is achieved through charring behaviour, calculated using advanced modelling to replace prescriptive fireproofing.
- Phase-Change Materials (PCM):
  - Embedded in walls or ceilings to store and release thermal energy, enhancing passive temperature regulation.
  - Ideal for stabilising internal environments in educational or office settings.
- Carbon-Negative Concrete and Geopolymers:
  - Used to reduce the embodied carbon footprint of structural elements.
  - Often selected with support from Life Cycle Assessment (LCA) tools and Environmental Product Declarations (EPDs).
- Smart Insulation and Aerogels:
  - Ultra-thin insulation systems that deliver exceptional thermal resistance, enabling compact or retrofit applications in constrained spaces.

The architect's responsibility is to evaluate new materials within the building's performance framework, ensuring compatibility with structural, fire, acoustic, and environmental criteria.

Summary: Innovation with Accountability

Flexibility in PBBD is not a design indulgence, it is a design responsibility. Architects are free to innovate, but only within the bounds of demonstrable performance. This shift requires:

- A solid grounding in engineering logic and environmental science,
- A fluency in simulation, testing, and iterative feedback,
- And a commitment to designing not just buildings that look inspiring, but ones that perform measurably better, for people, for the planet, and across the building's full lifecycle.

In this context, the architect becomes not just a creator of form but a strategic thinker, applying creativity through the lens of evidence, adaptability, and resilience.

Regulatory and Stakeholder Navigation

In performance-based building design (PBBD), the architect takes on a heightened responsibility both as a compliance strategist and as a communicator of performance intent. Because PBBD deliberately departs from rigid prescriptive pathways, the architect must guide the project through a more flexible, but often more complex, regulatory environment. This includes proving compliance through performance modelling and expert justification, securing buy-in from diverse regulatory and client stakeholders, and ensuring that all voices are heard and understood in a design process grounded in innovation and evidence.

The architect's role in this space is not merely to secure approvals but to establish credibility, manage risk, and build consensus across a broad range of interests, public, private, and technical.

Engaging Approval Bodies

One of the most critical tasks for architects in PBBD is to lead regulatory engagement from the outset, recognising that divergence from prescriptive codes often requires increased scrutiny and negotiation.

- Early engagement is key. Architects initiate dialogue with:
  - Planning authorities, to discuss massing, access, visual impact, and sustainability intent.
  - Building Control bodies, who review life safety, fire protection, egress, structure, and accessibility.
  - Fire and rescue authorities, especially in projects involving novel fire strategies, high-risk occupancies, or innovative materials.
  - Specialist regulators, such as Historic England, for heritage sites; the Health and Safety Executive (HSE), for high-rise residential projects (under the Building Safety Act 2022); and BREEAM or WELL assessors, for sustainability certifications.
- Performance-based approval strategies:
  - Instead of showing compliance with clauses of the Building Regulations (e.g., Approved Document B or L), the architect presents alternative solutions justified by simulation, expert endorsement, and risk assessment.
  - For example, where stair width or travel distance exceeds prescriptive limits, the architect collaborates with fire engineers to demonstrate equivalent or superior safety using tools such as evacuation modelling (RSET vs ASET) and smoke control CFD analysis.

  - In energy design, where a building does not follow the elemental standards of Part L (e.g., wall U-values or glazing ratios), the architect presents a whole-building energy model demonstrating compliance with carbon emissions benchmarks.
- Negotiation and iteration:
  - Regulatory feedback may prompt design adjustments, additional studies, or third-party peer reviews. The architect facilitates these responses, managing iterative submissions while ensuring design integrity is preserved.

The ability to speak both technical and regulatory language is critical, translating design intent into terms that satisfy the regulatory framework while championing the unique value of the performance-based approach.

Authoring and Curating Documentation

Documentation is the vehicle through which performance-based compliance is communicated, assessed, and approved. The architect plays a central role in both producing and curating the body of documentation that substantiates the performance claims of the design. Key documents may include:

- Performance-Based Fire Strategies:
  - Developed with fire engineers, these outline key safety concepts (e.g., evacuation time, fire growth scenarios, suppression strategy) and present the modelling evidence supporting deviations from prescriptive guidance.
  - May include zone diagrams, RSET/ASET analysis, smoke layer heights, and refuge strategies.
- Energy and Sustainability Reports:
  - These may include SAP/SBEM results for regulatory compliance, dynamic simulation results from IESVE or EnergyPlus, and narratives explaining net-zero, BREEAM, or WELL pathways.
  - Architects often contribute heavily to the architectural elements, façade specifications, window-to-wall ratios, passive design features, and ensure that MEP and structural data is appropriately integrated.
- Access and Inclusive Design Statements:
  - Especially in PBBD, inclusive design may deviate from prescriptive standards (e.g., using horizontal evacuation instead of vertical egress for mobility-impaired users). The architect must explain how inclusive outcomes are being achieved through performance, not just code compliance.

- Coordinated Drawing Sets:
  - Plans, sections, and details must reflect performance intent, showing fire compartments, ventilation pathways, escape routes, shading strategies, and material layer build-ups that align with modelling assumptions.
  - These are typically exported from BIM environments where spatial coordination and performance geometry are maintained.

The architect ensures that these documents tell a coherent, technically sound, and visually clear story, one that allows regulators and reviewers to understand how the proposed design meets or exceeds performance targets.

Stakeholder Communication and Visualisation

PBBD often involves sophisticated engineering strategies that must be made intelligible to non-technical stakeholders, including clients, user groups, funders, and community representatives. The architect serves as a translator, making complexity accessible through clarity and storytelling.

Key communication tools and strategies include:

- 3D visualisations and BIM models:
  - Used to illustrate spatial relationships, system integration, and circulation routes.
  - Can include time-lapse or animated sequences showing fire growth, smoke migration, or thermal gain across the day or year.
- Simulation Outputs:
  - CFD models, egress timelines, daylight distribution plots, and energy use graphs are annotated and embedded within presentations and reports to highlight key conclusions, not just raw data.
- Infographics and Diagrams:
  - These simplify layered technical systems into digestible overviews (e.g., showing the hierarchy of passive and active fire systems, or mapping environmental strategies by building zone).
- Scenario Walkthroughs:
  - Storyboarding worst-case fire events, peak summer temperature events, or system failures to show how the building performs under pressure, reassuring stakeholders of robustness and resilience.

This visual communication is not superficial, it is essential in aligning understanding, securing buy-in, and ensuring that performance decisions remain transparent and defensible throughout the project lifecycle.

Building Trust Through Transparency

At the heart of PBBD is the idea that performance outcomes matter more than box-ticking, but this flexibility depends on trust between the design team, the client, regulators, and the public. The architect plays a key role in building and sustaining that trust by ensuring:

- **Clarity in design rationale:** Stakeholders understand why a non-standard approach has been taken, and how it is supported by evidence and expert judgment.
- **Openness in process:** All simulations, calculations, and assumptions are clearly documented and made available for review and challenge.
- **Commitment to validation:** The performance intent is carried through to commissioning, testing, and post-occupancy evaluation. Architects may participate in soft landings or post-occupancy monitoring programmes to confirm that performance targets are being met in use.
- **Ethical responsibility:** In the context of life safety, climate change, and sustainability, architects are seen not only as designers but as custodians of public interest. A transparent and accountable approach to PBBD reflects that ethical duty.

Conclusion: Performance Advocacy as Architectural Leadership

In PBBD, the architect becomes not only a designer of space but an advocate of performance, leading the project through regulatory scrutiny, public accountability, and technical complexity. By engaging early with regulators, curating rigorous documentation, communicating clearly with stakeholders, and maintaining transparency, architects make performance-based innovation not only possible but credible.

This role demands confidence, diplomacy, and technical literacy—qualities that position the architect as a leader of performance outcomes and a trusted guide through the evolving landscape of high-performance building design.

RIBA Plan of Work Stages in PBBD Context

RIBA Stage	Architect's PBBD-Focused Role
Stage 0: Strategic Definition	Understand client vision; define project brief focused on performance outcomes rather than fixed solutions.
Stage 1: Preparation and Brief	Develop performance-based design brief with measurable goals (e.g., BREEAM rating, fire safety targets); collaborate on early risk and feasibility studies.
Stage 2: Concept Design	Translate performance objectives into spatial concepts; align massing, orientation, and materiality with performance metrics (daylight, energy, fire zones).
Stage 3: Spatial Coordination	Coordinate detailed inputs from fire engineers, M&E, and structure; ensure design allows for systems integration; develop geometry for simulation.
Stage 4: Technical Design	Finalise the design with all systems integrated; ensure drawings/specs reflect performance intent; support validation modelling (CFD, energy).
Stage 5: Manufacturing and Construction	Respond to contractor queries, conduct site inspections to verify that construction aligns with performance-driven specifications.
Stage 6: Handover	Assist in performance commissioning (POE, energy monitoring); contribute to post-occupancy evaluations and design tuning.
Stage 7: Use	Support building in operation with feedback and ongoing design input for lifecycle performance improvements.

Overview of the RIBA Plan of Work in the Context of Performance-Based Building Design

The RIBA Plan of Work provides a structured framework for managing building projects from inception to completion and beyond. Within the PBBD context, each stage becomes an opportunity to progressively define, test, validate, and realise measurable performance outcomes. Unlike traditional compliance-based workflows, PBBD demands a continuous feedback loop between design intent, performance metrics, and stakeholder engagement.

Stage 0: Strategic Definition

At this initial phase, the architect helps establish the fundamental purpose, scope, and ambitions of the project, laying the groundwork for a performance-led approach.

- Client engagement is focused on high-level aspirations: net-zero targets, climate resilience, operational efficiency, user well-being., or business continuity.
- Feasibility studies may include early site analysis, risk assessments (e.g., flood, seismic, fire), and strategic carbon impact assessments.
- Architects work with consultants to suggest a performance strategy framework that aligns with the client's priorities, regulatory obligations, and site context.
- Pre-brief questions around adaptive reuse, modular construction, or lifecycle value set the tone for innovation and flexibility.

In PBBD, Stage 0 is not about solutions but about framing the right questions: What does success look like? What must the building do, under normal and extreme conditions, to be considered high-performing?

Stage 1: Preparation and Brief

This stage is critical in shaping a robust performance brief, the foundation of all downstream design decisions and modelling.

- The architect leads stakeholder engagement (clients, user groups, facilities managers, sustainability advisors) to define specific and measurable performance goals, thermal comfort, daylight autonomy, fire safety benchmarks, energy intensity, acoustic targets, etc.
- Collaboration begins with fire engineers, MEP consultants, and sustainability specialists to define input parameters for simulations and testing.

- Risk matrices are developed to prioritise goals and identify key challenges (e.g., achieving ASET > RSET, designing for overheating under UK Climate Projections 2050).
- The architect translates these goals into design constraints, spatial requirements, and modelling needs, laying the groundwork for integrated performance testing at later stages.

The output of Stage 1 is not just a design brief but a performance specification, a strategic document outlining the intent and metrics that will drive all design decisions.

Stage 2: Concept Design

Here, the architect develops initial design proposals, massing, layout, and form, informed by the performance brief. This is the first major design phase where PBBD diverges from traditional workflows.

- Massing and orientation are evaluated for solar access, shading potential, and prevailing winds.
- Passive design strategies (e.g., natural ventilation, thermal mass, stack effects) are incorporated and tested for feasibility.
- Early digital models (BIM) are developed to a level sufficient for:
  - Daylighting simulations (e.g., DA, UDI)
  - Thermal comfort modelling (e.g., operative temperature, PMV)
  - Fire egress scenarios
- The architect tests and refines form based on initial simulation feedback, making concept design inherently performance responsive.

In PBBD, Stage 2 is where design intent meets measurable outcomes. Instead of focusing solely on aesthetics or function, architects begin shaping space to meet quantified performance targets.

Stage 3: Spatial Coordination

Also known as Developed Design, this stage is where spatial, technical, and regulatory alignment is consolidated, and where most performance validation begins in earnest.

- The architect coordinates inputs from the full consultant team into a spatially integrated model:
  - Structural grids and bracing systems
  - Compartmentation and fire strategy layouts

- Ductwork and risers for mechanical systems
- Zoning for HVAC, lighting, and automation

- CFD, dynamic thermal, evacuation, acoustic, and structural simulations are performed on updated models, with results used to inform refinements.
- The architect resolves clashes and contradictions between performance domains, such as balancing fire separation with spatial openness or reconciling acoustic and ventilation needs.

Stage 3 in PBBD is highly iterative. The architect ensures that the design is not just coordinated geometrically but also validated in terms of predicted performance, positioning the project for regulatory approval and internal sign-off.

Stage 4: Technical Design

At this stage, the design is taken to a level of technical completeness that allows construction to proceed, procurement to begin, and systems to be integrated in detail.

- The architect works with consultants and suppliers to embed performance into specifications, including:
  - Fire resistance ratings for elements
  - Glazing specifications (U-values, g-values, VT%)
  - Air permeability levels
  - Acoustic absorption and transmission properties
- BIM models are developed to full detail for performance-critical assemblies:
  - Thermal bridges eliminated or mitigated
  - HVAC coordination finalised with spatial tolerances
  - Structural performance under fire or seismic loads integrated into detailing
- Performance assumptions from earlier stages are checked against the detailed design, and simulations are rerun to confirm alignment.

For PBBD, Stage 4 is where design intent becomes design assurance. It's the point where performance no longer lives in theoretical models but is encoded in the drawings, schedules, and specifications issued for construction.

Stage 5: Manufacturing and Construction

During construction, the architect plays a vital role in ensuring that performance objectives are not lost in translation.

- Design intent is reviewed on-site to verify construction matches specifications, especially for:
    - Airtightness detailing
    - Fire compartmentation
    - Structural continuity
    - System installation (ductwork, sprinklers, insulation)
- The architect may attend or witness commissioning tests, including:
    - Blower door tests for air leakage
    - Smoke control system testing
    - Lighting lux level verification
    - HVAC balance and air change rate measurements
- Value engineering pressures during procurement are carefully managed to avoid compromising performance goals.
- Design changes are reviewed and modelled (where necessary) to confirm continued compliance.

In PBBD, Stage 5 isn't just about "building what's on the drawings", it's about building what performs. The architect becomes a key guardian of fidelity between simulated and as-built outcomes.

Stage 6: Handover

At project completion, the building is handed over to the client, but in PBBD, this is also the start of performance verification.

- The architect assists in producing performance manuals that explain how systems are intended to operate in real use.
- Soft landings processes may be led or supported by the architect, providing guidance to facilities teams in system controls and expected behaviours.
- Where contractual or regulatory performance metrics exist (e.g., BREEAM credits, WELL certification, EPC targets), the architect helps document and verify that they have been achieved.

This is a critical time for aligning real-world use with design intent, ensuring that the building operates as promised, and that lessons are captured for future design improvement.

Stage 7: Use

In PBBD, the building's lifecycle doesn't end at completion, it continues through ongoing monitoring, evaluation, and learning.

- The architect may support or lead Post-Occupancy Evaluation (POE), gathering data on energy use, IAQ, occupant comfort, and operational issues.
- User feedback is collected and assessed against original performance objectives, closing the loop between design, simulation, construction, and experience.
- If performance falls short, the architect may coordinate design or system changes, particularly in buildings with ongoing commissioning or energy performance contracts.

Stage 7 reflects the true spirit of PBBD: designing for performance in use, not just in theory. It positions the architect as a lifelong partner in the building's success.

Conclusion

The RIBA Plan of Work provides a powerful framework for project delivery, but in the context of PBBD, it becomes a roadmap for evidence-based design, validation, and continuous improvement. The architect's role expands across all stages to ensure that spatial quality, functional utility, and measurable performance are tightly interwoven, from concept through occupancy.

Plan Production in Performance-Based Building Design

In Performance-Based Building Design (PBBD), architectural drawings and documentation are not just instruments of communication, they are tools of validation, evidence of compliance, and records of performance intent. As the project progresses through the RIBA stages, the architect's responsibility for producing and managing documentation becomes increasingly technical, interdisciplinary, and simulation driven.

Unlike traditional prescriptive design, where drawings primarily reflect dimensional accuracy and regulatory compliance, PBBD demands that drawings serve as interfaces between design and performance modelling, enabling analysis, simulation, and regulatory approval. Each drawing or model must be performance-intelligent, containing the level of detail, logic, and clarity required for engineering interpretation and testing.

Performance Briefs and Conceptual Diagrams

At early stages (RIBA 1–2), the architect develops drawings that illustrate performance strategies rather than finalised geometries. These conceptual outputs help articulate how the proposed design responds to performance objectives and serve as the first communication layer with clients and consultants.

Typical early-stage plan production includes:

- **Solar Path Diagrams:** Showing building orientation, sun angles by season, and the relationship between façade design and solar exposure. These inform passive heating and daylighting strategies.
- **Wind Flow and Ventilation Diagrams:** Using site analysis and prevailing wind data to map potential for natural ventilation, cross-ventilation corridors, and air intake/exhaust points.
- **Fire and Egress Diagrams:** Indicating occupant loads, exit routes, stair locations, refuge areas, and travel distances in relation to fire compartments and critical risk zones.
- **Climate Response Sketches:** Illustrating strategies such as shading, green roofs, thermal zoning, or night flushing based on the local climate profile.

These diagrams are not contractual, but they are strategic, used to initiate simulation modelling, guide stakeholder dialogue, and form the foundation of the performance brief.

Design Coordination Drawings

As the design progresses into RIBA Stage 3 (Spatial Coordination), the architect produces increasingly detailed plans, sections, and details that synthesize inputs from multiple disciplines. In PBBD, these coordination drawings are essential in resolving performance domains within spatial and structural frameworks.

Key outputs include:

- **Fire Strategy Plans:**
    - Showing fire zones, protected lobbies, compartmentation lines, smoke barriers, detection and suppression systems.
    - Coordinated with evacuation modelling and fire engineer input.
- **HVAC and Mechanical Integration Plans:**
    - Reflecting duct runs, plant rooms, ceiling voids, risers, vents, and louvres—all coordinated with structural and spatial design.
    - Thermal zoning and air change strategies are overlaid to reflect ventilation performance assumptions.
- **Structural Coordination:**
    - Structural grids, transfer beams, and load paths are shown in plan and section, linked to the architectural layout and system access needs.

  - Coordination ensures that fire and acoustic requirements (e.g., slab penetrations, intumescent coatings) are accommodated.
- **System Overlay Drawings:**
  - Multi-layered overlays showing the spatial relationships and routing of passive and active systems (e.g., sprinkler networks, smoke vents, photovoltaics, thermal envelopes).

In PBBD, these drawings serve not only for coordination, but also as source material for validation modelling, and must reflect reality with fidelity and clarity.

Model-Ready Geometry

One of the defining characteristics of PBBD is the use of simulation tools to model performance, thermal, structural, acoustic, fire, daylight, etc. For this to be possible, architectural geometry must be prepared in a way that is compatible with analytical software.

The architect's responsibilities include:

- BIM Model Structuring:
  - Developing the model with clear zoning, separable layers (walls, floors, ceilings, systems), and data-rich components.
  - Assigning materials with correct thermal, acoustic, and fire properties (linked to manufacturers or environmental data sets).
- Exporting to Analytical Formats:
  - Ensuring models are clean, consistent, and exportable to platforms like IESVE, EnergyPlus, Ladybug Tools (for Rhino/Grasshopper), FDS (Fire Dynamics Simulator), or Pathfinder (for egress).
  - Avoiding common pitfalls such as overlapping geometry, missing joins, or non-parametric elements.
- Simulation Input Collaboration:
  - Providing geometries that match the assumptions set out in the performance brief and coordinating with engineers to validate that boundary conditions and occupancy profiles are accurate.

Without simulation-ready geometry, PBBD cannot progress beyond intent. The architect plays a vital role in bridging the creative and the computational, ensuring that spatial design can be interrogated through engineering logic.

4. Detailed Plans and Specifications

By RIBA Stage 4 (Technical Design), architectural documentation becomes increasingly prescriptive, used to direct construction and embed performance parameters. These outputs must fully describe how the building's systems will work, where they are located, and how they will be constructed.

Core responsibilities include:

- **Detailing for Performance:**
  - Wall build-ups showing insulation types, fire barriers, acoustic separations, and airtight layers.
  - Junctions (floor-to-wall, wall-to-roof, etc.) resolved to prevent cold bridging, air leakage, or structural conflict.
  - Firestopping and compartmentation details coordinated with both fire strategy and acoustic/thermal performance.
- **Schedules and Specifications:**
  - Windows and doors schedules with U-values, SHGC, sound attenuation, and fire ratings.
  - Product specifications linked to environmental declarations (EPDs), fire test certificates, or acoustic ratings.
- **Integrated Drawings:**
  - Clear labelling of performance zones (e.g., smoke reservoirs, protected routes, ventilation zones).
  - Use of legends and tags to reference model components back to performance criteria.
- **Code and Performance Referencing:**
  - Annotated references to performance criteria rather than just compliance clauses (e.g., "Façade to meet average U-value < 1.0 W/m^{2}K and g-value < 0.3 in cooling zone").

At this stage, documentation becomes a contractual representation of performance; accuracy, clarity, and completeness are critical.

5. Validation and Compliance Packages

To obtain regulatory approval or third-party certification (e.g., BREEAM, LEED, WELL, fire authority approval), the architect must coordinate and/or contribute to the production of performance validation packages, bundles of documents that demonstrate how the proposed or constructed building satisfies the agreed performance metrics.

These packages typically include:

- Performance Summary Reports:
  - A narrative summarising how each design element contributes to performance outcomes, cross-referencing drawings, simulations, and consultant inputs.
- Simulation Outputs and Diagrams:
  - Charts, graphs, and heat maps from tools such as:
  - Daylight distribution (DA, UDI)
  - Energy consumption and breakdown (EUI, HVAC demand)
  - Fire/smoke spread animations
  - Egress flow maps and clearance times
- As-Built Reconciliations (in later stages):
  - Updates to drawings and models to reflect construction changes.
  - Photographic records or site verification logs of key performance features (e.g., insulation installation, firestopping).
- Regulatory Crosswalks:
  - Documents mapping how alternative strategies meet or exceed the intent of prescriptive regulations (e.g., fire stair capacity, air quality thresholds, acoustic ratings).

These validation packages serve as the evidence base for sign-off, and the architect ensures they are complete, coherent, and align with both the original brief and the final product.

Conclusion

Plan production in PBBD is a multi-layered process that evolves from strategic sketching to rigorous technical detailing, with each stage reinforcing the building's ability to perform in use, under stress, and over time. The architect plays a pivotal role not just as a producer of drawings, but as a curator of performance evidence, connecting space, systems, simulations, and standards into a unified, buildable, and testable whole.

Summary

The Architect as a Performance Integrator in PBBD

In the era of Performance-Based Building Design (PBBD), the architect is no longer just the originator of form or the aesthetic leader of the project; they are a strategic integrator, orchestrating the convergence of design vision, technical performance, regulatory compliance, and end-user needs. This role is at once dynamic, technical, and creative, demanding fluency in modelling and metrics alongside empathy, judgement, and imagination.

A Holistic, Multidisciplinary Role

The performance-based approach elevates the architect's position within the design team. Architects are no longer confined to responding to pre-set rules; instead, they take on the responsibility of defining bespoke performance targets with clients and consultants and ensuring those targets are met through every stage of the design and construction process. Whether the goal is operational energy reduction, occupant safety, adaptive reuse, or zero-carbon design, the architect becomes the translator of high-level objectives into buildable, measurable outcomes.

From Vision to Validation

Unlike prescriptive design, where rules define the outcome, PBBD invites the architect to begin with outcomes and work backward, shaping the design to deliver them. This calls for a continuous loop of ideation, simulation, coordination, and adjustment. The architect is required to work hand-in-hand with engineers, specialists, and clients to explore design options, test them against performance models, and refine them based on evidence. It's a process where intuition is tested by data, and form is shaped by function and resilience.

Through model-ready geometry, performance mapping, and simulation-informed layouts, architects ensure that each line on a drawing is not just spatially correct but performance-intelligent, ready to be tested, validated, and ultimately trusted by stakeholders and regulators.

Performance Across the RIBA Stages

The architect's influence in PBBD extends across the full arc of the RIBA Plan of Work:

- At Stage 0–1, they help define performance ambitions and risk appetites, shaping the brief through strategic dialogue.
- In Stages 2–3, they synthesise technical inputs and drive spatial coordination to create performance-integrated designs.

- During Stage 4, they convert design intent into construction-ready drawings that embed critical performance layers—fire, thermal, acoustic, environmental.
- In Stages 5–6, they oversee construction quality and help commission systems, ensuring that what is built aligns with what was modelled.
- Finally, in Stage 7, they support post-occupancy evaluation and feedback, closing the loop between projected and real-world performance.

Designing for Use, Not Just Compliance

PBBD shifts the architectural focus away from minimal code compliance and toward optimising building behaviour in use. This means designing not only for first occupation but for long-term adaptability, resilience, and well-being. Architects in PBBD must consider lifecycle emissions, occupant experience, operational energy, and even failure scenarios. They must ask: How does this space respond in a heatwave? In a fire? Over 60 years of changing use?

This forward-thinking, evidence-based mindset demands that architects embrace uncertainty, model possibilities, and design for robustness. Their work is not only about creating spaces that delight but about delivering environments that endure.

Communicating Performance to All Stakeholders

Throughout the process, the architect is also a storyteller of performance, presenting complex technical strategies to non-technical audiences. Whether through BIM visualisations, scenario diagrams, or validation packages, the architect must bridge the gap between simulation and understanding, between engineering logic and human needs.

In PBBD, performance is not just calculated, it is communicated, negotiated, and made real through drawings, dialogue, and shared vision.

Conclusion: The Architect as Custodian of Performance

Ultimately, PBBD redefines architectural practice not as the pursuit of formal expression alone, but as the custodianship of performance. Architects are the guardians of intent, ensuring that the design not only looks good on paper but performs as promised in practice, now, and into the future.

By embracing this expanded, interdisciplinary role, architects become the connective tissue between aspiration and action, between systems and space, and between data and design. In doing so, they shape not only better buildings, but better futures.

Chapter 4

Fire Engineers in Performance-Based Building Design

Designing Bespoke Fire Strategies for Complex and High-Risk Structures

Introduction

In Performance-Based Building Design (PBBD), fire engineers play a pivotal role by developing customized fire safety strategies that go far beyond prescriptive code compliance. Particularly in complex or high-risk buildings, such as high-rises, transport hubs, hospitals, or heritage sites, prescriptive methods may be too rigid, simplistic, or unfit for unique layouts or uses. Performance-based fire engineering empowers designers to craft tailored solutions rooted in science, simulation, and risk-informed analysis.

The Need for Bespoke Fire Strategies

Conventional building codes typically assume standard layouts, occupancies, and hazards. However, modern buildings increasingly deviate from such norms. Large atria, interconnected floors, mixed-use spaces, and unusual geometries present challenges that cannot be fully addressed through prescriptive methods alone.

Fire engineers in PBBD respond by creating fire strategies that are specifically tailored to the building's:

- Geometry and spatial configuration
- Occupancy type and density
- Operational use and activities
- Fire load distribution and compartmentation
- Structural resilience and evacuation provisions

Key Components of a Performance-Based Fire Strategy

In Performance-Based Building Design (PBBD), fire safety is no longer achieved solely through prescriptive code compliance. Instead, it relies on evidence-based strategies that integrate fire science, structural response, human behaviour, and system performance. Fire engineers, in collaboration with structural and M&E teams, develop strategies that predict how a building will perform under fire and ensure that occupants can safely evacuate while the structure remains stable long enough to support life safety and firefighting operations.

A critical focus of this strategy lies in:

- Maintaining structural integrity during fire exposure
- Ensuring smoke and heat do not compromise evacuation routes
- Matching escape times with the available safe period before conditions become untenable
- Coordinating detection, suppression, and control systems to support performance outcomes

Key Components of a Performance-Based Fire Strategy

A robust fire strategy in PBBD is supported by four interdependent modelling domains. Each contributes to a holistic understanding of fire development, impact, and mitigation—allowing for tailored solutions that balance architectural ambition, life safety, and cost efficiency.

Fire Growth Modelling

This component focuses on how a fire ignites and develops over time in a given space. It provides the foundation for all subsequent modelling domains, since the characteristics of fire growth (e.g., rate of heat release, smoke production) drive structural loading, visibility, and tenability conditions.

Key Parameters:

- Heat Release Rate (HRR): Measured in kW, it defines the fire's size and growth rate
- Ignition delay and flashover thresholds
- Material combustion properties: calorific value, flame spread index, charring rate
- Ventilation-limited vs fuel-limited fires

Application:

- Defines design fire scenarios, both most likely and worst-case, based on contents, configuration, and ventilation

- Allows engineers to design fire suppression, ventilation, and detection systems tailored to expected fire dynamics
- Informs thermal loading inputs for structural fire engineering analysis

Tools:

- Smartfire, FDS (Fire Dynamics Simulator) for 3D fire development
- CFAST, B-Risk for zone-based thermal stratification and smoke layer prediction

PBBD Insight: Realistic fire growth modelling supports performance justification for non-standard materials, open-plan layouts, and innovative compartmentation strategies.

Smoke Movement Modelling

Smoke is the primary life safety hazard in fires—responsible for reduced visibility, respiratory harm, and panic. Understanding its movement is critical to:

- Designing egress routes
- Protecting vertical shafts (stairs, lifts)
- Controlling smoke in atria or interconnected volumes

Key Inputs:

- Airflow patterns from HVAC systems
- Wind pressures and buoyancy-driven flows
- Leakages through doors, penetrations, and façade interfaces
- Thermal plumes and layer stratification

Design Outcomes:

- Pressurised stairwells to keep escape routes clear
- Smoke curtains and exhaust vents in large open areas
- Smoke reservoirs and fire curtains for atria or mezzanine floors
- Airflow direction control at escape routes or refuge areas

Tools:

- FDS, PyroSim, Smartfire for full CFD simulation of smoke travel and tenability (visibility, CO levels, temperature)
- CFAST for zone-based calculations in compartmentalised buildings

PBBD Application: Modelling allows for justified alternatives to prescriptive smoke barriers or compartmentation, supporting design creativity without compromising safety.

Evacuation Modelling and Human Behaviour Analysis

The cornerstone of fire strategy is ensuring that all occupants can reach safety before conditions become untenable. PBBD allows engineers to simulate occupant behaviour and movement, accounting for building complexity, demographics, and emergency procedures.

Key Concepts:

- **Required Safe Egress Time (RSET):** Time needed for all occupants to evacuate
- **Available Safe Egress Time (ASET):** Time before conditions (smoke, heat, toxicity) become life-threatening
- **Pre-movement time:** Delay between alarm and movement (varies by use, alertness, training)
- **Travel speed:** Adjusted for age, mobility, crowd density, smoke visibility

Design Applications:

- Validates stair width, travel distances, and number of exits
- Supports refuge floor planning and progressive evacuation strategies
- Optimises exit signage placement, auditory/visual alarms, and staffed evacuation protocols
- Ensures inclusion of occupants with mobility or cognitive impairments

Tools:

- Pathfinder, BuildingExodus, MassMotion, Simulex
- Often integrated with fire and smoke simulation outputs for coupled analysis

PBBD Standard: Demonstrate that RSET < ASET with at least a 100–120 second safety buffer, depending on occupancy and fire load.

Suppression System Effectiveness

Active systems are modelled not only for compliance, but for their quantifiable impact on fire development and tenability.

Suppression Types:

- **Sprinklers (wet, dry, pre-action):** Delay or control flame spread and cool the space
- **Water mist systems:** Provide cooling and oxygen displacement in smaller droplets—ideal in heritage or data centres
- **Gaseous suppression:** Inert or chemical agents (e.g., FM200, NOVEC 1230) for electrical and high-value spaces
- **Foam systems:** Used in hangars, fuel storage, or process facilities

Modelling Use:

- Simulate time to activation based on RTI (Response Time Index), ceiling height, and detector spacing
- Adjust fire growth curves to reflect suppression impact
- Inform structural fire resistance ratings by demonstrating reduced thermal exposure
- Coordinate suppression logic with BMS, fire alarm, HVAC shutdown, and emergency lighting

Tools and Data Sources:

- Fire modelling tools with suppression algorithms (FDS + PyroSim)
- Empirical data from FM Global, NFPA, manufacturer curves
- BS7974: Framework for the application of fire engineering principles to the design of buildings

PBBD Advantage: Enables strategic trade-offs e.g., permitting extended travel distances or reduced fire resistance levels based on early suppression and detection.

Structural Resilience Integration

Fire can compromise structural stability via:

- Thermal weakening of steel or spalling of concrete
- Loss of cross-section in timber or unprotected steel
- Uneven expansion and load redistribution leading to disproportionate collapse

Structural Fire Engineering in PBBD:

- Conducts thermal-structural FEA simulations to predict deformation and failure modes
- Designs for load redistribution in case of localised failure

- Includes robustness and redundancy: multiple load paths, protected elements, and connections designed to tolerate fire-induced movement
- Evaluates structural performance for safe evacuation duration and firefighting access
- Standards Referenced: EN 1991-1-2 (Eurocode fire actions), BS 9999, NFPA 5000, ASCE/SEI guidelines

A Performance-Based Fire Strategy as a Multi-System Model

PBBD fire strategies are no longer static compliance documents—they are living performance frameworks, tested through digital models and scenario analysis. Fire engineers integrate:

Domain	Goal	PBBD Outcome
Fire Growth	Define fire intensity and duration	Thermal loading for structure and suppression
Smoke Movement	Map spread and visibility impact	Design of clear escape routes
Evacuation Modelling	Simulate occupant safety	Confirm ASET > RSET across conditions
Suppression Systems	Control or delay fire	Reduce risk, permit alternative designs

Together with structural and M&E teams, fire engineers ensure that buildings not only comply but perform under stress, protecting life, limiting damage, and maintaining structural integrity until full evacuation is complete.

Facilities for the Fire Service and Firefighter Safety in PBBD Fire Strategies

In Performance-Based Building Design (PBBD), a robust fire strategy must not only account for occupant evacuation and structural resilience but also provide for firefighter operations and safety. Buildings must support the timely, effective, and safe intervention of firefighting personnel, particularly in complex, high-rise, mixed-use, and high-risk structures.

Fire service facilities are not simply add-ons; they are integral elements of the fire strategy that influence:

- The location and layout of access routes and firefighting shafts
- Compartmentation and suppression zoning

- Smoke control and ventilation interface
- Fire detection and communication systems
- Structural robustness during firefighting operations

PBBD enhances traditional fire service provisions through risk-informed planning, simulation modelling, and scenario testing, enabling design teams to justify tailored provisions while ensuring firefighter safety and operational efficiency.

Core Objectives of Fire Service Provisions in PBBD

1. Enable safe, rapid entry and access to all parts of the building.
2. Provide protected firefighting routes and zones that withstand heat and smoke.
3. Ensure critical systems (e.g. lifts, water supply, communications) are resilient and operable during fire events.
4. Support way finding, ventilation control, and spatial awareness for crews in low-visibility, high-stress conditions.
5. Minimise structural and environmental hazards to firefighters during prolonged intervention.

Key Components of Fire Service Facilities

Firefighting Access and Approach Routes

Effective fire response begins at the site perimeter. PBBD incorporates:

- **Fire appliance hard standings:** Sized and strengthened for turning radii and weight of modern fire vehicles.
- **Site access roads:** Compliant with gradient, clearance, and proximity standards.
- **Building orientation and signage:** Clear way finding for fire crews, including fire control room signage, fire hydrant markers, and zone ID.

Design Reference: BS7974, Approved Document B (ADB), BS 9999, and local Fire and Rescue Service operational guidance.

Firefighting Shafts and Stairs

These are vertically protected routes that provide smoke-free access for firefighters to upper or lower floors.

PBBD Enhancements:

- Pressurisation modelling to test smoke control effectiveness under worst-case stack effect and door-opening conditions.
- Increased stair width or intermediate landings to support two-way movement and equipment staging.
- Thermal protection to stair cores, verified through CFD-based thermal modelling.
- Real-time signage integrated with BMS to guide firefighters to the correct shaft or stair during multi-compartment fires.

Performance Consideration: Stairs should maintain tenable conditions for at least 60–120 minutes, depending on risk classification and height

Firefighting Lifts and Evacuation Lifts

Where floor levels exceed 18 m or for complex vertical access, firefighting lifts are essential.

Key PBBD Considerations:

- CFD and fire growth modelling used to validate lobby protection, shaft pressurisation, and lift pit drainage.
- Elevator call overrides tested in simulated scenarios.
- Dual power supply and fire-rated control systems evaluated for failure risk and redundancy.
- In some cases, combined use as evacuation lifts may be permitted, subject to performance-based justification.

Target Performance: Lift lobby temperature < 60°C, visibility > 10 m, smoke ingress prevented for defined time under peak fire load.

Fire Control Rooms and Firefighter Communication Systems

These serve as **information hubs** for incoming fire crews. They must be:

- Easily accessible from the main entrance
- Physically protected (e.g., 120-minute rated enclosure)
- Equipped with:
    - Building floor plans and fire strategy drawings
    - Fire alarm panel and zone indicators
    - CCTV, sprinkler flow indicators, and HVAC status

- Manual ventilation override and pressurisation control
- Battery-backed lighting and power supply

Firefighter Communication:

- Firefighter telephones or wireless radio repeater systems across all floors
- Wayfinding light paths to assist crews in low-visibility conditions
- Public Address and Voice Alarm (PAVA) override for emergency comms

PBBD Addition: Fire control systems can be linked to digital twin dashboards, allowing real-time thermal and smoke layer awareness during intervention.

Fire Hydrant Systems and Internal Water Supplies

Internal firefighting systems must be tailored to risk level, height, and building use.

Systems Include:

- Wet and dry risers for floor levels above 18 m
- Hose reels in industrial or high-hazard zones
- Sprinkler and deluge systems interfaced with manual override
- Landing valves and branch pipe access within protected lobbies

PBBD Validation:

- Flow and pressure simulations ensure adequate coverage at top floor under simultaneous demand.
- Zone isolation testing supports phased firefighting and suppression during system maintenance or sectional shutdown.

Performance Goal: Maintain minimum 8 l/sec at 4 bar at the highest outlet for effective firefighting operations.

Structural Integrity During Firefighting

Firefighting operations impose unique mechanical, thermal, and dynamic loads on structures:

- Concentrated water loads from suppression (e.g., deluge)
- Dynamic loads from movement of personnel and equipment
- Prolonged thermal exposure during mop-up and overhaul stages

Structural Fire Engineering Considerations:

- FEA-based structural stability models test performance under prolonged fire beyond standard exposure curves.
- Collapse prevention objectives defined for key access routes and staging zones
- Protected zones (e.g., fire staging areas, protected lobbies) kept structurally intact for 2–4 hours

Reference Standards: EN 1991-1-2, BS 7974, and local fire service resilience criteria.

Designing for the Fire Service in PBBD

Performance-Based Fire Strategy extends far beyond occupant evacuation. It must enable, support, and safeguard the fire crews tasked with extinguishing the fire and protecting life and property.

M&E, architectural, and structural designers must work with fire engineers to:

- Understand the operational needs and limitations of fire service personnel
- Simulate realistic scenarios where fire spread, smoke migration, and crew movement converge
- Design buildings that don't just protect people—but enable people to protect buildings

Firefighter safety is not optional, it's designed, simulated, and verified.

The chart below summarises the above discussion.

Applications in Complex or High-Risk Structures

High-Rise Buildings

High-rise buildings are defined under UK guidance (Approved Document B) as buildings exceeding 18 metres in height, and under BS 9991/9999 and the Fire Safety (England) Regulations 2022, even more stringent measures are applied above 30 metres. These buildings pose amplified fire safety challenges due to their scale, vertical circulation systems, compartmentation complexity, and occupant load. Prescriptive codes often fall short in addressing these challenges holistically, making Performance-Based Fire Engineering (PBFE) essential

Why PBFE Is Essential in High-Rise Design

Traditional code-based approaches assume predictable egress, conventional compartmentation, and symmetrical building geometry. However, high-rise buildings often include:

- Unconventional layouts (e.g., offset cores, sky gardens, open internal atria)
- Mixed-use occupancy (e.g., commercial, residential, leisure)
- Large numbers of occupants spread across multiple vertical levels
- Reliance on mechanical systems (e.g., fire lifts, pressurisation, smoke extraction)

These conditions demand a risk-informed, scenario-tested, performance-driven fire strategy developed with advanced modelling and interdisciplinary collaboration.

Key Fire Engineering Considerations

Vertical Fire and Smoke Spread

- In high-rise structures, chimney effects, stack effects, and wind-driven smoke movement significantly impact smoke behaviour.
- Service risers, lift shafts, and façade cavities can act as vertical smoke conduits if unprotected.
- CFD Modelling (e.g., via Smartfire, FDS or PyroSim) is used to simulate:
    - How smoke will spread floor-to-floor
    - The effect of opening or failing doors
    - Smoke entrainment into stairwells and lift lobbies
    - Cross-ventilation between compartments via leaks or forced ventilation

These simulations help to design robust smoke containment and extraction systems, especially where open-plan or interconnected floors are present.

Pressurised Stairwells and Escape Shafts

- Pressurisation systems are critical to maintaining clear escape routes.
- PBFE defines:
  - Optimal air supply rates to create positive pressure against smoke ingress
  - Pressure differential tolerances for door opening force (e.g., <100 N per BS EN 12101-6)
  - Modelling of air leakage, shaft volume, and door swing impacts under fire conditions
  - Impact of external wind pressure at high elevations (wind stack effect)

Performance models are stress-tested for system reliability under worst-case ventilation failure, stairwell door openings, and elevator shaft breaches.

Phased Evacuation Strategies

Full evacuation is often not practical or safe in tall buildings due to:

- Congested stairs
- Extended evacuation times
- Potentially counterproductive re-entry into smoke-laden floors

Instead, phased evacuation involves:

- Immediate evacuation of fire-origin floors
- Sequential evacuation of adjacent floors based on fire escalation modelling
- Use of public address systems, smart alarms, and command-and-control protocols

Evacuation Modelling (e.g., using BuildingExodus, Pathfinder, MassMotion):

- Maps crowd movement, densities, bottlenecks, and behavioural variations
- Validates that Required Safe Egress Time (RSET) is less than Available Safe Egress Time (ASET) with adequate margins
- Simulates partial stair blockage, fatigue delay, and alternative route selections

Refuge Floors and Areas of Relative Safety

- In high-rise towers, not all occupants may be able to descend stairs, particularly those with reduced mobility, panic, or injury.
- Refuge floors or protected lobbies act as temporary holding spaces and must:

- Have minimum 60–120 minutes fire resistance
- Be kept free of smoke via pressurisation or smoke lobbies
- Be directly accessible from escape cores or lifts
- Contain emergency communication systems for disabled users and rescue teams
- Be sized according to the expected occupant load and evacuation timeline

PBFE evaluates:

- The tenability conditions inside refuge spaces using temperature, CO, O_2 levels, and visibility metrics
- The resilience of communication and alarm systems over long durations (e.g., >30 minutes before rescue)

Integrated Suppression and Detection Systems

- High-rise fire strategies often feature multi-layered protection, including:
  - Residential sprinklers (per BS 9251 or BS EN 12845)
  - Smoke extraction or containment in plant areas and basements
  - Early detection via aspirating smoke detectors or addressable systems

PBFE evaluates:

- The response time of systems in various ignition scenarios
- Their synergy with HVAC, pressurisation, and fire alarm logic
- System redundancy in the event of power or pump failure

Case Example: Performance-Led Stair Core Optimisation

Project: 40-Storey Mixed-Use Tower in Central London

Challenge: Architect proposed a design with only two stair cores to maximise usable floor area. Prescriptive codes required three.

Solution:

- Evacuation modelling showed both stair cores could accommodate full egress within required time frames under simultaneous evacuation scenarios.
- CFD modelling validated pressurisation kept stair cores smoke-free for >60 minutes under worst-case fire.
- Compartmentation and automated voice alert systems supported phased evacuation.

- The refuge areas on every fifth floor were modelled to maintain tenable conditions for >90 minutes.

Outcome: Approved by the authority having jurisdiction based on a robust fire engineering justification package, allowing architectural efficiency without compromising safety

Regulatory Interface

- Strategies are typically aligned with:
  - BS 9991 / BS 9999 – Code of practice for fire safety in residential and non-residential buildings
  - BS7974: Framework for the application of fire engineering principles to the design of buildings
  - Fire Safety (England) Regulations 2022 – For external wall risks and evacuation plans
  - Building Safety Act 2022 – For higher-risk residential buildings (HRRBs)

Performance-based fire strategies are submitted with:

- Fire Statement
- Evacuation Strategy
- CFD and egress modelling reports
- Risk assessments for non-compliant features

Conclusion

High-rise buildings demand an advanced and flexible approach to fire safety, one that balances regulatory compliance, occupant protection, and architectural innovation. Performance-Based Fire Engineering offers the tools to simulate, analyse, and justify bespoke strategies that address the unique vertical, spatial, and human factors of tall buildings. Through advanced modelling, detailed risk assessment, and integrated system design, fire engineers ensure that even the most ambitious high-rise structures remain safe, resilient, and future-ready.

Transport Hubs and Airports

Transport hubs, including airports, railway stations, bus terminals, and multi-modal interchanges, present some of the most complex challenges for fire safety engineers. These buildings combine vast open volumes, fluctuating occupancies, and dynamic usage patterns with critical life safety, security, and operational continuity objectives. Many of these structures operate 24/7 and house a wide mix of transient users, staff, and vulnerable individuals.

Performance-Based Fire Engineering (PBFE) provides a vital methodology for achieving fire safety objectives without compromising the architectural and operational ambitions of these unique facilities.

Key Considerations

Large Atrium Spaces and Terminal Volumes

Airports and major stations often incorporate grand atria or expansive terminal halls, designed for visual openness, way finding, and flexibility. However, these spaces introduce unique fire dynamics challenges:

- *Smoke Stratification:* In very tall spaces (often 20m+), smoke can form stratified layers below the ceiling, making it difficult to predict smoke movement and visibility at occupant level.
- *Reservoir Sizing:* PBFE helps define smoke layer interface height to ensure sufficient tenability (visibility >10m, temperature <60°C, CO concentration <1,400 ppm).
- *Extraction and Make-up Air:* Advanced Computational Fluid Dynamics (CFD) modelling (e.g., via Smartfire, FDS or PyroSim) is essential to design:
    - Smoke extract rates (typically in m^3/s per fire size)
    - Make-up air inlets and their impact on smoke flow
    - Buoyancy-driven and forced ventilation interaction

Design Example:

A 200,000 m^3 terminal atrium may require multiple smoke zones with partial barriers, fan systems, and ducts sized to extract >150 m^3/s while ensuring the smoke layer remains >2.5m above floor level for evacuation visibility.

Delayed and Dynamic Evacuation

Evacuation in transport hubs is notably non-linear and non-predictable, owing to:

- **Unfamiliar Users:** Most occupants are transient travellers unfamiliar with the layout, making signage and auditory systems more critical.
- **Language and Cognitive Barriers:** Multilingual communication strategies are required for alarms and PA announcements.

- **Encumbered Movement:** People carrying bags, pushing trolleys, or travelling with children significantly slow flow rates.
- **Sleeping or Distracted Occupants:** Lounges, cafes, and waiting zones may have seated or sleeping individuals with delayed reactions.

Performance Modelling Approaches:

- Evacuation Simulations (using BuildingExodus, Pathfinder, MassMotion) consider:
  - Extended reaction times (30–120 seconds for sleeping users)
  - Delayed egress start times
  - Use of alternative exits if primary routes are obstructed
- Egress route effectiveness is tested under both normal and low visibility conditions, incorporating detours, blocked escalators, or failed systems.

Key Metric:

Maintaining an RSET (Required Safe Egress Time) below the ASET (Available Safe Egress Time) with a ≥2-minute safety margin under worst-case assumptions is typical in PBFE justifications for terminal spaces.

Multi-Alarm and Sequential Fire Scenarios

In large transport facilities, the possibility of simultaneous or sequential fire incidents in different areas is a real risk:

- Fires in baggage handling areas, retail zones, or mechanical plant rooms can trigger different systems independently.
- Scenario planning must consider alarm zoning, how and when different sections of the terminal initiate evacuation.
- Evacuation modelling must verify that:
  - Movement from one zone does not conflict with another.
  - Egress routes do not overlap dangerously.
  - Occupants rerouted from a primary exit have a secondary path with sufficient capacity.

PBFE tools simulate multi-point ignition events, especially in high-occupancy conditions such as holiday travel periods or major event surges.

Redundancy Planning:

- Backup extraction fans
- Battery-supported emergency lighting
- Cross-zonal pressurisation
- Alternate communication systems (e.g., digital signage and beacons)

Interfacing Emergency Interlocks with M&E Systems

Fire engineers collaborate closely with mechanical, electrical, and controls engineers to ensure proper system interlocks and real-time response coordination. These integrated systems include:

- **HVAC Shutdown or Diversion:** Ensuring mechanical ventilation doesn't exacerbate smoke spread.
- **Lift Protocols:** Firefighting lifts activate while passenger lifts are decommissioned or directed to safe zones.
- **PA System Integration:** Automatically switches from general announcements to fire alerts, including multilingual and zone-specific messages.
- **Digital Wayfinding Displays:** Linked to fire systems to indicate safe exits or blocked paths.
- **Platform Safety Systems:** In underground or rail-linked hubs, coordination with signalling systems, tunnel ventilation, and platform screen doors is essential.

Control Logic Diagrams (often in BIM and IFC format) are tested through fire engineering analysis to confirm proper response sequencing and timing.

Example: International Airport Atrium Smoke Strategy

Context: A new UK international airport terminal featured a central atrium spanning four floors, with adjacent departure gates, lounges, and retail.

Challenge: The architectural vision required an open plan with full visibility across levels—making traditional full-height fire curtains or barriers unfeasible.

PBFE Approach:

- CFD models simulated 3 MW and 6 MW fire scenarios at peak and off-peak hours.
- Smoke reservoirs were created using discreet partial barriers (1.5–2.0m in height) and curved geometry aligned with the architecture.

- Jet fans and high-level ducted extraction provided dynamic smoke removal above the occupancy plane.
- Evacuation modelling confirmed tenability on all levels for >8 minutes post-ignition, with total evacuation achieved in <6.5 minutes.
- PA system integration delivered zoned, time-delayed messaging to optimise phased egress and prevent crowding.

Outcome: The design received approval without imposing visual barriers, preserving openness, and improving passenger flow, backed by a robust, scenario-tested fire strategy.

Standards and Guidance Referenced

- BS 9999: Code of practice for fire safety in commercial buildings
- BS7974: Framework for the application of fire engineering principles to the design of buildings
- NFPA 130: Fixed guideway transit and passenger rail systems
- ICAO Annex 14: Airport safety provisions
- BS EN 12101-5/6/13: Smoke and heat control systems
- CIBSE Guide E: Fire safety engineering

Conclusion

In transport hubs and airport terminals, Performance-Based Fire Engineering provides the only practical and adaptable method of achieving safe egress, smoke management, and operational continuity. Through advanced modelling of occupant behaviour, fire dynamics, and system interlocks, engineers ensure that even the most visually ambitious and operationally complex transport facilities deliver life safety without compromising user experience or architectural integrity.

Hospitals and Care Homes

Hospitals, care homes, and other healthcare facilities represent some of the most challenging environments for fire safety design. These are buildings where life safety cannot rely solely on rapid self-evacuation, because many occupants may be:

- Bed-bound
- Non-ambulatory or mobility-impaired
- Cognitively impaired (e.g., dementia patients)
- Connected to life-support systems or undergoing surgery

Performance-Based Fire Engineering (PBFE) offers a vital approach to tailor fire strategies around clinical realities, occupant vulnerability, and staff-assisted evacuation procedures, while maintaining compliance with regulations and continuity of care.

Key Fire Safety Considerations in Healthcare Settings

Progressive Horizontal Evacuation (PHE)

Unlike most buildings that use vertical egress strategies, hospitals and care homes typically rely on progressive horizontal evacuation, a method where patients are moved laterally into adjacent fire compartments rather than evacuating the building.

Core Principles:

- **Compartmentation:** Buildings are divided into fire-resistant zones (typically 30 or 60-minute compartments), allowing phased evacuation without full building clearance.
- **Adjacent Safety:** Each compartment must be large enough to accommodate occupants from a neighbouring compartment in an emergency.
- **Time Gained:** PHE is designed to buy time until either the fire is suppressed, or full evacuation becomes necessary.

Engineering Contributions:

- Fire resistance analysis to justify wall, ceiling, and floor performance under expected fire loads.
- Time-equivalent exposure modelling to assess if 30–60-minute compartmentation is sufficient based on expected fire growth and suppression time.
- Tenability Modelling: CFD simulations assess whether safe conditions (e.g., temperature < 60°C, CO < 1,400 ppm, visibility > 10m) are maintained in adjacent compartments for 15–30 minutes.

Standards Referenced:

- HTM 05-03 Part B (UK Department of Health): Fire precautions in new healthcare buildings
- BS 9999: For vertical egress and evacuation lift integration
- BS7974: Framework for the application of fire engineering principles to the design of buildings

Staff-Assisted Movement and Occupant Dependencies

Patients in healthcare settings cannot self-evacuate. Their safe movement depends entirely on staff availability, equipment, and procedural readiness.

Modelling Factors:

- **Caregiver-to-patient ratios** (e.g., 1:3 or 1:5 during night shifts)
- **Evacuation aids:** Beds, wheelchairs, evacuation sleds, or transfer boards
- **Transfer times:** Moving a single bed-bound patient can take 2–5 minutes
- **Congestion impacts:** Corridor widths, turning radii, and fire door delays affect speed

Simulation Tools:

- BuildingExodus, Pathfinder or MassMotion adapted for hospital conditions:
- Assign mobility types (e.g., ambulatory, assisted, bed-bound)
- Incorporate staff allocation algorithms
- Model staged evacuation across compartments and floors

Shift-Based Scenario Planning:

- Night-time modelling (e.g., 1 nurse per 10 patients) must demonstrate evacuation feasibility using reduced human resources.
- Fire drills and scenario tests verify if predicted timelines match operational capacity.

Metric Example:

A stroke ward may require 18–25 minutes to move all patients from one 30-minute compartment to another. PBFE must verify that this can be achieved before ASET thresholds are reached

Smoke and Fire Compartmentation Integrity

Compartment integrity is critical in healthcare environments where evacuation delays are expected, and patients may remain in place for extended periods.

Engineering Considerations:

- **High Fire Loads:** Imaging suites (MRI, CT), storage rooms, and treatment areas often exceed average fire loads due to electronics, gases, and chemicals.

- **Medical Gases:** Oxygen, nitrous oxide, and hydrogen-rich environments increase flame propagation potential.
- **Smoke Movement:** Pressurisation systems and CFD simulations evaluate smoke leakage through ducts, door gaps, and voids.
- **Duct Penetrations:** AHUs, return air grilles, and nurse call systems are tested for fire dampers or fire-stopping.

Testing Protocols:

- Fire engineers review manufacturer data and apply fire-resisting element calculations (per BS EN 1366, BS 476 Part 20-22).
- CFD visualisation confirms smoke cannot breach critical care or neonatal wards within ASET timelines.

Barrier Example:

Fire-rated glazed partitions with automatic drop-down screens may be used in reception or triage zones to balance visibility with protection.

Redundant Fire Protection Systems

Given the high consequence of failure, healthcare fire strategies rely on multiple layers of redundancy.

Systems Typically Deployed:

- Wet-pipe sprinkler systems to BS 9251 for early suppression
- Gaseous suppression (e.g., FM-200 or inert gas) for IT rooms and surgical theatres
- Double-detection logic (e.g., smoke + heat) to reduce false alarms and unwanted patient movement
- Evacuation lifts certified under BS EN 81-72 or BS 9999 to aid vertical patient movement
- Back-up generators and fire panel failover systems

Modelling Tasks:

- Suppression effectiveness simulations to determine how quickly the fire size stabilises
- Response time analysis: Time from ignition to suppression, to ASET breach, to full evacuation

- Redundancy mapping to confirm that even under single-point failure (e.g., loss of mains power), critical systems remain operational

Insight:

Performance-based strategies can compensate for limited egress routes by demonstrating superior early suppression and compartmentation—provided this is backed by modelling and risk assessment.

Example: Acute Hospital PBFE Scenario

Project: A new-build 500-bed acute hospital in England

Challenge: Evacuation time for a 36-bed ward during night staffing hours

Methodology:

- Fire initiated in patient room with highest fire load
- Evacuation simulations run with 1 nurse per 12 patients
- Patients divided into:
  - 40% bed-bound
  - 30% assisted ambulatory
  - 30% independent walkers
- Compartmentation modelled for 60-minute resistance
- CFD analysis simulated tenable conditions in the corridor and adjacent compartments

Outcome:

- 90% of patients evacuated in <30 minutes under conservative assumptions
- Additional control measures recommended:
  - Use of lightweight evacuation trolleys
  - Integration of emergency lighting triggered by smoke alarms
  - Enhanced staff training on low-staff evacuation protocols

Guidance and Standards Referenced

- HTM 05-01 and HTM 05-03: Fire code for healthcare premises
- BS 9999: Code of practice for fire safety in the design and management of buildings
- NFPA 101 – Chapter 18 & 19: Life Safety Code for healthcare occupancies
- BS EN 81-72: Firefighting lifts in buildings

Conclusion

Hospitals and care homes cannot rely on traditional fire strategies that assume rapid and unaided evacuation. Through Performance-Based Fire Engineering, fire strategies are custom-designed around real-world clinical workflows, occupant vulnerabilities, and operational constraints. Using advanced modelling tools, compartmentation analysis, and system redundancy design, PBFE delivers tailored solutions that balance patient safety, regulatory compliance, and continuity of care.

Heritage and Listed Buildings

Heritage and listed buildings are treasured for their cultural, architectural, and historical significance, but this legacy often comes with complex fire safety challenges. Many of these buildings were constructed long before modern fire codes existed, using materials and methods that are inherently combustible or lack structural compartmentation.

Design interventions must tread a careful line between preserving historic character and delivering robust fire protection. This makes them ideal candidates for performance-based fire engineering, where bespoke, non-intrusive strategies can be modelled, validated, and justified—ensuring life safety without compromising conservation goals.

Core Challenges in Heritage Fire Safety

- **Lack of Original Compartmentation:** Many historic buildings feature open-plan layouts, grand staircases, or timber-framed construction with no inherent fire separation.
- **Combustible Materials:** Common materials include lath-and-plaster ceilings, timber panelling, thatched roofs, or early glues and finishes with high flammability.
- **Non-Compliant Egress:** Escape routes may be narrow, circuitous, or terminate in internal courtyards or upper levels with no secondary egress.
- **Installation Constraints:** Drilling into protected stonework, plaster mouldings, or panelling for M&E services (e.g., sprinklers, alarms) may be prohibited under heritage regulations.
- **Irregular Visitor Flow:** Museums, galleries, and public heritage sites often have visitors unfamiliar with the layout, which may include multi-level transitions and limited way finding.

Key Considerations in Performance-Based Design for Heritage Buildings

Discreet Fire Suppression Systems

Conventional sprinkler systems may damage historic finishes or require unsightly piping that conflicts with conservation principles. Performance-based strategies introduce low-visibility, low-water-volume solutions, such as:

- **Water Mist Systems:** Fine water droplets (typically <100 microns) suppress fires with reduced water damage risk. These systems are ideal for artefact-rich spaces like libraries, churches, or archives.
- **Concealed Sprinklers:** Recessed, colour-matched or pop-down sprinkler heads that blend with ceilings or cornices.
- **Inert Gas Systems:** In confined areas (e.g., document vaults, historic server rooms), inert gases like argon or nitrogen displace oxygen to suppress fires without water use.
- **Hybrid Systems:** Systems combining mist and gas, especially useful for semi-enclosed gallery spaces.

Engineering Validation: PBFE practitioners model suppression effectiveness using fire growth simulations (e.g., via Smartfire) to demonstrate that the reduced water discharge or agent concentration still achieves sufficient flame knock-down within the Available Safe Egress Time (ASET).

Advanced Fire Detection and Alarm Systems

Sensitive, early detection is vital in heritage buildings, especially where extended evacuation or fire service response times may be needed. Invasive alarm installations may not be permitted, so alternatives include:

- **Aspirating Smoke Detection (ASD):** Uses capillary tubing hidden in cornices or conduits to draw air samples continuously to a central sensor. These systems detect minute smoke levels before visible flames occur.
- **Wireless Sensors:** Ideal for listed interiors where cable runs cannot be embedded. Devices are networked and battery-powered, with radio signal redundancy.
- **Beam Detectors:** Infrared or laser beams across large open volumes (e.g., chapels, banqueting halls) detect obscuration from smoke without ceiling-mounted units.

- **Multilingual PA Systems and Sounder Beacons:** Particularly in public heritage spaces, PBFE integrates these with emergency voice communication to assist way finding and guidance.

Integration with Modelling: Detection times are factored into fire growth curves to verify early alarm activation and correlate with suppression system timing and RSET calculations.

Behavioural Evacuation Planning and Public Wayfinding

Evacuation in heritage sites presents a behavioural complexity uncommon in conventional buildings:

- Visitors are often unfamiliar with routes, especially in maze-like castles or stately homes with multiple staircases, mezzanines, or courtyards.
- Visitors may hesitate, pausing to collect belongings or reluctant to leave valuable displays behind.
- Egress routes are often uneven or historic, with narrow, single-exit staircases or external fire escapes unsuitable for high-volume discharge.
- PBFE addresses these factors using:
    - Crowd modelling tools (e.g., BuildingExodus, Pathfinder, MassMotion) to simulate real-time egress, factoring in crowd hesitancy, visual confusion, and mixed mobility levels.
    - Enhanced way finding design, including:
    - Discreet directional LED floor lighting
    - Glow-in-the-dark historic signage
    - Portable steward communication tools
    - Assisted evacuation protocols for guided visitor movement by trained staff or docents.

Simulation Example:

In a World Heritage cathedral, models showed that voice-guided phased evacuation combined with discreet perimeter signage achieved full visitor clearance 3 minutes faster than alarm-only strategies.

Fire-Resisting Upgrades to Historic Structures

Structural modifications in heritage buildings are often heavily restricted. However, fire engineers can specify non-intrusive passive solutions to improve compartmentation and fire resistance:

- **Intumescent Coatings:** Transparent or colour-matched finishes applied to timber beams, wall panelling, or joinery to provide 30–60 minutes of protection without altering visual character.
- **Fire Curtains and Drop Screens:** Concealed within ceilings or beams and automatically deployed in fire events, these barriers assist in smoke containment and zone separation.
- **Glazing Upgrades:** Historic windows and fanlights can be retrofitted with laminated fire-resisting glass or secondary glazing that meets BS EN 1364.
- **Floor Strengthening:** Timber floors and staircases are underlaid with mineral wool or fire boards, hidden beneath floorboards or tapestries.

Performance Analysis: Engineers model fire load densities and run time-equivalent analyses to demonstrate how these upgrades delay heat transfer and allow for safe evacuation before structural failure.

Example: Grade I Listed Museum Fire Strategy

Project: A Grade I listed museum in a converted manor house, featuring open galleries, original timber floors, and priceless artefacts.

Challenges:

- No existing compartmentation or mechanical extraction
- Single protected stairwell for three floors
- Restrictions on visible fire doors or sprinkler heads

PBFE Solutions:

- CFD smoke simulations demonstrated that early detection via aspirating sensors provided enough warning time for all occupants to evacuate before tenability limits were breached.
- Water mist suppression in mechanical plant areas and kitchen zones was validated to prevent flashover.
- Drop-down smoke curtains were installed above archways, concealed within architectural detailing.
- Compartmental time modelling confirmed that fire would remain confined to the origin room for >20 minutes, sufficient to clear the floor via the existing stairwell.

Outcome: The local authority accepted the performance-based fire strategy in lieu of full compliance with compartmentation requirements, preserving the architectural integrity of the listed building.

Benefits of Performance-Based Fire Engineering in Heritage Contexts

Flexibility

- Enables conservation-compliant solutions tailored to building-specific constraints.
- Permits architectural freedom to retain vaulted ceilings, open staircases, and interconnected galleries.

Optimisation

- Resources are directed toward high-risk areas (e.g., kitchens, electrical rooms) while preserving low-risk historic zones.
- Avoids unnecessary fireproofing of low-load spaces based on risk and fire dynamics modelling.

Safety Assurance

- Modelling validates safe evacuation timelines under worst-case ignition and crowd scenarios.
- Realistic fire scenarios account for late detection, material combustion properties, and visitor behaviours.

Integration

- Coordinates fire detection and suppression systems with architectural heritage and M&E constraints.
- Enables integration with digital twin technologies and BIM overlays for fire risk and conservation monitoring.

Future-proofing and Adaptability

- Supports adaptive reuse, turning heritage sites into venues, museums, or education centres with compliant fire strategies.
- Allows updates through post-occupancy review and remote monitoring of system performance.

Conclusion

Heritage and listed buildings require a delicate and informed approach to fire safety, one that honours the past while protecting life in the present. Performance-Based Fire Engineering offers the technical agility to design site-specific, non-intrusive fire strategies that meet both conservation and safety mandates. By combining simulation, passive upgrades, discreet technology, and behavioural modelling, fire engineers play a transformative role in ensuring that historic architecture remains not only admired, but also resilient, accessible, and safe.

The Fire Engineer as Fire Guardian

Custodian of Fire Strategy Integrity

- In performance-based building design, where prescriptive code pathways may be exceeded or adapted, the fire engineer must ensure that the rationale behind the fire safety design is preserved and followed through during implementation.
- They champion the original design assumptions, such as evacuation modelling, smoke control systems, and structural fire resistance, ensuring they are not compromised during value engineering or construction.

Interface Between Design and Occupation

- A fire guardian works across disciplines—coordinating with architects, M&E engineers, structural engineers, and authorities—to maintain fire safety principles during design and technical detailing.
- During construction, they provide oversight, reviewing changes, substitutions, or site conditions that could impact compartmentation, egress, detection, suppression, or passive protection systems.
- In post-occupancy, they may support commissioning, handover, and maintenance protocols, ensuring the systems perform as modelled.

3. Advocate for Risk-Based, Scenario-Led Design

- In PBBD, the fire engineer develops and validates performance-based solutions (e.g., alternative egress strategies, smoke control modelling, phased evacuation) that differ from code minimums.
- The fire guardian ensures that these scenarios are understood by all stakeholders, not just accepted on paper but prepared for in practice.

4. Educational and Operational Resource

- Particularly in complex or high-risk environments (hospitals, airports, high-rise buildings), the fire guardian may train facilities teams, occupants, and safety officers in operational fire safety, including:
    - The logic behind the evacuation plan,
    - The performance of fire doors, lobbies, or smoke extraction,
    - The implications of system failure or misuse.

5. Lifecycle Accountability

- The role doesn't end at handover. In modern regulatory environments (e.g., under the UK's Building Safety Act 2022), there's growing emphasis on the "golden thread" of information, requiring ongoing record-keeping and traceability of fire safety features.
- A fire guardian may assist in maintaining that thread, supporting in-use reviews, building safety case reports, and long-term compliance.

Why It Matters in PBBD

In PBBD, where performance strategies often push the envelope of standard practice, the fire guardian provides continuity and reassurance. They ensure that:

- Assumptions in evacuation modelling match real-world operation,
- Smoke control systems are maintained and are functional,
- Compartmentation is not breached by late-stage service changes,
- All building users understand and can act on the strategy in place.

This concept aligns with emerging industry best practices and regulatory frameworks that emphasise competence, continuity, and accountability in fire safety—especially post-Grenfell and within the UK's new Building Safety Regime.

Chapter Summary: Fire Engineering in Performance-Based Building Design

Fire engineering is a cornerstone of performance-based building design (PBBD), providing a scientific and risk-informed framework for protecting life, property, and business continuity in ways that go beyond prescriptive code compliance. This chapter explored the evolution, principles, and practical applications of fire engineering in complex and high-performance buildings, emphasising the architect's collaboration with fire engineers and the critical role of coordinated, evidence-based design strategies.

Key Themes Covered

From Compliance to Performance

Traditional fire safety design relies heavily on prescriptive rules, e.g., fixed stair widths, maximum travel distances, or compartment sizes. PBBD departs from this model by allowing alternative strategies to be used where performance can be proven. This has opened the door to more flexible, innovative, and efficient solutions, particularly in large-scale or non-standard buildings (e.g., high-rises, mixed-use developments, atria, and heritage retrofits).

Core Tools and Methodologies

The chapter introduced core methodologies used in fire engineering, including:

- Fire modelling using Computational Fluid Dynamics (CFD) to simulate heat and smoke behaviour.
- Evacuation modelling to assess safe egress time (RSET) relative to available safe egress time (ASET).
- Compartmentation design and smoke control strategies tailored to specific scenarios.
- Risk assessments and scenario planning as central components of performance justification.

Architectural Integration

Fire safety is not an isolated discipline. The chapter explored how fire engineering must be embedded in:

- Spatial planning (e.g., locating escape routes and refuge areas),
- Structural design (e.g., maintaining integrity under fire loading),
- Services coordination (e.g., avoiding breaches in fire-rated construction).

Architects work in tandem with fire engineers to test layouts, materials, and access strategies, ensuring that aesthetic and functional goals are compatible with fire safety.

Lifecycle Considerations and Adaptive Safety

In PBBD, fire safety does not stop at design sign-off. Buildings must be designed to perform over time, as occupancies change, systems age, and new technologies emerge. This requires fire safety strategies to be robust, adaptable, and clearly documented, forming part of the building's operational and maintenance lifecycle.

The Fire Guardian: A Key Role in PBBD

A central concept introduced in this chapter is the fire engineer's evolving role as the fire guardian, a professional custodian of the building's fire safety integrity from conception through occupation and beyond.

Key Functions of the Fire Guardian

- **Design Integrity:** Ensuring the assumptions behind performance-based strategies (e.g., egress time, smoke control, structural resistance) are followed throughout detailing, procurement, and construction.
- **Construction Oversight:** Monitoring key installations (e.g., fire-stopping, sprinkler layout) to prevent late-stage deviations that could compromise safety.

- **Operational Interface:** Supporting building managers with training, handover documentation, and commissioning, so that systems are used and maintained as intended.
- **Lifecycle Safety:** Contributing to the "golden thread" of safety information, especially under the UK's post-Grenfell building safety regime, by providing consistent, traceable advice across design, delivery, and occupancy phases.

By assuming the role of fire guardian, the fire engineer becomes a continuous presence in the building's safety journey, helping to maintain accountability and ensure that performance strategies deliver as promised in both normal and emergency conditions.

Conclusion

Fire engineering in PBBD represents a fusion of science, creativity, and responsibility. By moving beyond rigid codes, it enables smarter, more site-specific solutions, but demands rigorous modelling, stakeholder trust, and ongoing verification. The collaboration between architect and fire engineer, anchored by clear communication, simulation-driven design, and the guiding hand of the fire guardian, ensures that fire safety is not just designed, but continuously delivered.

This chapter underlines that performance-based fire safety is not just about escaping fire, it is about designing buildings that are resilient, adaptable, and inherently safe by design.

Chapter 5

Structural Engineers in Performance-Based Building Design

In Performance-Based Building Design (PBBD), structural engineers move beyond code-driven sizing and generic safety margins. They play a pivotal role in ensuring that buildings are not only code-compliant under normal conditions but are also resilient, adaptive, and safe under extreme or unexpected structural demands.

Through the use of advanced computational modelling, probabilistic analysis, and risk-based design, structural engineers develop bespoke structural strategies tailored to a building's use, location, performance objectives, and potential exposure to hazards.

Role of Structural Engineers in PBBD

Traditional prescriptive codes often limit innovation and assume "average" scenarios. PBBD empowers structural engineers to:

- Customise structural solutions to meet specific performance targets (e.g., immediate occupancy after an earthquake).
- Justify non-traditional materials or forms using advanced simulations.
- Optimise structural design for cost, safety, durability, and sustainability without over-engineering.

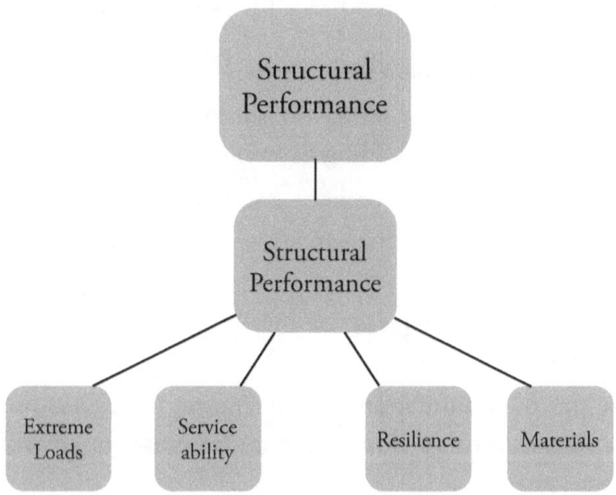

Their responsibilities span across:

1. Extreme Load Event Modelling
2. Serviceability and Comfort Design
3. Resilience and Redundancy Planning
4. Material Innovation and Lifecycle Thinking
5. Interdisciplinary and Digital Integration

Structural Performance Under Extreme Loads

Structural engineers in PBBD address rare but high-consequence scenarios with probabilistic and simulation-based approaches. These may include:

Seismic Performance

Although the UK is classified as a low-seismicity region, Performance-Based Seismic Design (PBSD) remains essential for certain building types and infrastructure. Critical assets such as nuclear facilities, hospitals, transport nodes, bridges, and tall or irregular structures require a seismic performance strategy that goes beyond minimum prescriptive compliance. Globally, PBSD is a cornerstone of resilient and adaptable building design, particularly in regions where seismic events are more frequent or severe (e.g., Japan, Turkey, Italy, New Zealand, California).

PBSD focuses on how a building behaves under seismic excitation, rather than merely ensuring it avoids collapse. This approach allows structural engineers to design buildings that:

- Remain operational after moderate earthquakes
- Protect life safety during rare, large events
- Recover quickly from damage, reducing downtime and cost

Why Seismic Performance Matters in PBBD

Prescriptive seismic codes are based on simplified assumptions about ground motion, material behaviour, and building dynamics. These codes often:

- Apply generic safety factors without evaluating the actual dynamic response
- Lack the ability to tailor performance to function (e.g., life safety vs. continued operation)
- Cannot account for irregular geometries, non-ductile elements, or base-isolated systems

PBBD overcomes these limitations by integrating detailed simulations and real-world ground motion data to develop structural systems that are tailored to project-specific performance goals.

Core Components of Performance-Based Seismic Design

Advanced Dynamic Analysis Techniques

Rather than static load equivalents, PBSD employs time-domain and frequency-domain analyses that model actual seismic ground shaking effects on the structure.

Key Methods:

- **Non-linear Time History Analysis (NLTHA):**
    - Uses real or synthetic earthquake records to simulate building response over time.
    - Captures non-linearities, plastic deformations, and energy dissipation mechanisms.
- **Pushover Analysis:**
    - Assesses progressive collapse by incrementally increasing lateral forces.
    - Identifies weak links, soft storeys, and plastic hinge formation points.
- **Incremental Dynamic Analysis (IDA):**
    - Tests structure against increasing ground motion intensity levels to evaluate collapse probability.

Software Commonly Used: OpenSees, SAP2000, ETABS, Perform-3D, ANSYS, Abaqus

Performance Objectives and Limit States

PBSD defines multiple performance tiers depending on building type, occupancy, and criticality. These performance objectives are commonly aligned with Eurocode 8, FEMA P-58, or ASCE 41.

Typical Limit States:

Performance Objective	Description	Target Occupancy
Operational (OP)	No structural or non-structural damage	Critical facilities
Immediate Occupancy (IO)	Minor damage: structure remains usable	Hospitals, data centres
Life Safety (LS)	Damage controlled; no collapse	Commercial/residential
Collapse Prevention (CP)	Severe damage, but no progressive failure	Minimum code level

Engineers must model and demonstrate that the structure meets these objectives under specified seismic intensity levels (e.g., 475-year or 2,475-year return periods)

Structural Strategies for Seismic Resilience

To achieve these performance goals, PBBD applies a combination of structural detailing, material strategies, and advanced systems:

Base Isolation Systems

- Decouples the structure from ground motion using elastomeric bearings, lead-rubber isolators, or sliding systems.
- Reduces acceleration transmitted to superstructure by up to 90%.
- Ideal for museums, hospitals, heritage buildings, or high-value infrastructure.

Energy Dissipation and Damping Devices

- Adds viscous dampers, friction dampers, or tuned mass dampers to reduce motion.
- Particularly useful in retrofit projects or long-span, flexible structures.
- Buckling-Restrained Braced Frames (BRBFs) dissipate energy through inelastic buckling in both directions.

Moment-Resisting Frames (MRFs)

Steel or reinforced concrete frames with rigid connections designed to yield under seismic load without brittle failure.

Provide ductility and rotation capacity, absorbing and redistributing energy.

Shear Walls and Core Systems

- Engineered concrete or steel walls resist lateral loads and limit drift.
- Strategically located around lifts, stairs, or services for architectural integration.

Modelling Output: Lateral drift (Δ), base shear (V), floor acceleration (a), residual deformation, hinge formation maps.

Regional Context – Seismic Considerations in the UK

While the UK is not a high-risk seismic zone, PBSD is still required in the following contexts:

- Nuclear power facilities (ONR and IAEA seismic classification)
- Transportation infrastructure (e.g., HS2 bridges, tunnels)

- Tall buildings with irregular torsional modes or high slenderness ratios
- Critical civil defence or energy assets
- Sites built on poor soils, former mining areas, or regions of localised seismic amplification (e.g., Cumbria)

PBSD allows designers to justify the omission or reconfiguration of certain features (e.g., expansion joints, mass dampers) based on real-time structural response rather than code prescription.

Example Application: Hospital on Soft Soil with Emergency Function

Scenario: A 12-storey urban hospital located in an area with low-to-moderate seismicity and soft clay subsoil.

PBSD Approach:

- Conducted site-specific response spectra analysis to understand ground amplification effects.
- Designed dual-system lateral load resistance: reinforced shear walls + damped braced steel frames.
- Targeted Immediate Occupancy performance objective.
- Applied non-linear time history analysis using 7 scaled ground motion records.
- Verified drift ratios <1.5%, no residual tilt, and non-structural acceleration <0.3g.

Outcome: Demonstrated that even under a 2,475-year return period event, the hospital would remain operational, satisfying national health continuity standards and emergency response capability.

Summary: Seismic Performance in PBBD

Seismic performance within Performance-Based Building Design is not limited to avoiding collapse, it is about predicting, modelling, and managing damage in a way that:

- Preserves life and property
- Supports continuity of operations
- Aligns design with use, value, and risk profile
- Enables architectural and material flexibility through structural validation

With tools like non-linear time history analysis, performance tier modelling, and advanced damping technologies, structural engineers design not for the average earthquake, but for the real, quantifiable demands of the worst-case scenario, and how the building must perform during and after it.

Wind and Storm Resistance

In Performance-Based Building Design (PBBD), wind and storm resistance is no longer limited to applying static wind loads from basic pressure charts. Instead, it involves advanced dynamic analysis, site-specific environmental modelling, and interdisciplinary collaboration to create buildings that are not only structurally safe, but also comfortable, resilient, and efficient in high-wind environments.

This is especially critical for:

- High-rise buildings that sway or vibrate under wind loading
- Coastal and offshore structures exposed to tropical cyclones or intense pressure gradients
- Bridges, stadiums, and long-span roofs, which are susceptible to aeroelastic phenomena like flutter and vortex shedding
- Slender towers or cantilevered forms, where structural stability may be compromised by wind-induced oscillations

Why Wind and Storm Resistance Matters in PBBD

While prescriptive codes (e.g., Eurocode EN 1991-1-4 or ASCE 7) provide basic wind pressure values, they often:

- Assume regular building forms and uniform terrain
- Do not address dynamic response (acceleration, resonance)
- Lack provisions for occupant comfort in flexible structures
- Ignore aeroelastic instabilities and wind-structure interaction

In PBBD, engineers adopt a site-specific, performance-tiered approach, combining analytical tools, physical modelling, and parametric design integration to ensure that wind effects are managed intelligently and proactively.

Advanced Methods in Wind Performance Engineering

Dynamic Wind Response Modelling

High-rise and long-span structures must be assessed for dynamic wind effects, not just peak pressure but also time-varying fluctuations that cause motion, discomfort, or structural fatigue.

Key Phenomena:

- **Vortex Shedding:** Alternating low-pressure vortices form on the leeward side of the structure, creating lateral oscillations at discrete frequencies (Strouhal number-driven).
- **Crosswind Oscillations:** Can cause swaying in a direction perpendicular to wind flow, especially problematic in tall, narrow buildings or open-framed structures.
- **Buffeting:** Wind gusts and turbulence causing irregular loading, especially at corners or gaps between buildings.
- **Flutter and Galloping:** Aerodynamic instabilities in lightweight structures such as cable-stayed bridges or tensile roof systems.

Key Metrics: Natural frequency (Hz), damping ratio (%), critical wind speed (m/s), base moment (kNm), tip deflection (mm)

Wind Tunnel Testing and CFD Integration

PBBD employs both physical testing and computational fluid dynamics (CFD) to simulate complex airflow behaviour around structures:

Wind Tunnel Testing

- Scaled models tested in boundary layer wind tunnels that replicate real wind profiles over terrain categories (e.g., urban, coastal, mountainous).
- Measures pressure taps, aerodynamic coefficients, vortex formations, and cladding load distributions.
- Essential for non-standard shapes, closely spaced towers, or buildings near steep topography.

CFD Simulations

- Engineers simulate full-scale wind flows in 3D, visualising velocity vectors, pressure zones, and flow separation.
- CFD tools like ANSYS Fluent, SimScale, or OpenFOAM allow parametric iteration and integration with architectural design.
- CFD results can be fed into finite element models (FEM) to analyse stress, deformation, and fatigue.

Design Outcomes: Improved building orientation, aerodynamic tapering, recessed balconies, and dynamic mass dampers for vibration control.

Occupant Comfort and Motion Criteria

In tall or flexible buildings, wind may not threaten structural integrity, but it affects usability, comfort, and perception of safety.

Comfort Performance is measured in:

- Horizontal and vertical acceleration thresholds for different building types (residential, hotel, office)
- UK and international standards (e.g., ISO 10137, CIBSE Guide A, ANSI A117)
- Maximum peak accelerations in the 10–20 milli-g range (mg) are usually tolerated before discomfort arises

Mitigation Strategies:

- **Tuned Mass Dampers (TMDs):** Large pendulum-like weights mounted at the top of towers to counteract motion.
- **Aerodynamic Modifications:** Varying roof heights, softening corners, introducing porous cladding or wind baffles.
- **Increased Damping:** Using structural connections or interior elements that increase energy dissipation.

Example: A 300m tower in the Gulf region reduced peak lateral acceleration by 30% through TMDs and aerodynamic corner recessing, verified via time-history wind excitation modelling.

Resilience Against Storms and Cyclonic Winds

For coastal buildings or those in typhoon/cyclone-prone areas, PBBD includes provisions for storm surge, uplift, and high-speed loading scenarios:

- **Windborne Debris Impact Modelling:** Glazing and façade systems are evaluated for penetration resistance using projectile tests (e.g., missile test standards from ASTM E1996 or EN 13049).
- **Roof Uplift Simulation:** CFD analysis models suction loads on large, low-slope roofs, critical in airport terminals, warehouses, or arenas.
- **Pressure Equalisation and Drainage Design:** Structures are shaped or vented to avoid internal pressurisation and water ingress during high-speed wind events.

Envelope Detailing: Integration of cyclonic-rated shutters, reinforced cladding fixings, and edge flashings to prevent uplift or cladding failure.

Example: High-Rise Residential Tower in a Wind-Intensive Region

Context: A 60-storey tower located in a coastal UK city exposed to prevailing westerlies.

PBBD Approach:

- Site-specific wind tunnel testing using a 1:300 scale model with surrounding topography.
- CFD simulations refined façade articulation and orientation for wind deflection.
- Vortex shedding frequency matched to structural natural frequency; TMD installed to reduce resonance risk.
- Occupant comfort modelling based on ISO 10137 showed accelerations remained <15 mg at 95th percentile conditions.
- Envelope tested for dynamic pressure resistance, preventing wind-induced façade vibration and glazing blowout.

Outcome: The final design reduced structural steel tonnage by 12% while exceeding both structural and comfort performance targets.

Standards and Guidance Referenced

- Eurocode EN 1991-1-4: Wind actions on structures
- ASCE 7-22: Minimum design loads including wind, snow, seismic
- ISO 10137: Evaluation of human response to building motion
- CIBSE Guide A: Environmental design
- NBCC and BS 6399 (for wind loading in UK and international contexts)

Summary: Wind and Storm Resistance in PBBD

Performance-Based Wind Engineering allows structures to respond not just to expected environmental conditions, but to real-world dynamic behaviours through:

- Site-specific modelling
- Vibration and comfort criteria
- Advanced tools like CFD and wind tunnels
- Interdisciplinary coordination with façade and architectural teams

PBBD enables safer, more efficient, and more comfortable structures, particularly in high-rise, long-span, and exposed environments. Wind resilience becomes a design driver, not a constraint.

Accidental and Impact Loads

- Engineers assess progressive collapse, designing against events like vehicular impact, blast, or structural member failure.
- Alternate load path analysis identifies whether loads can be redistributed in case of localized failure.
- Protective measures may include standoff distances, reinforced concrete walls, or post-tensioned frames.

Example: In airport terminals, structural engineers simulate impact from aircraft fuel explosions or vehicular ramming to validate structural survivability and controlled collapse mechanisms.

Serviceability, Comfort, and Long-Term Performance

While life safety during extreme events like earthquakes or hurricanes is critical, a structure's day-to-day performance under typical conditions and its serviceability are equally important in Performance Based Building Design (PBBD). Buildings must remain comfortable, functional, and durable throughout their design life. Serviceability issues, if unaddressed, can lead to user dissatisfaction, material damage, increased maintenance costs, or functional obsolescence long before structural failure is even a concern.

PBBD approaches serviceability not as a set of arbitrary limits but as a performance goal, quantified through analysis, simulation, and predictive modelling. Engineers work closely with architects and building operators to align technical performance with user expectations, especially in high-performance, long span, or sensitive environments.

Key Serviceability and Comfort Considerations

Deflection Limits and Floor Deformation

Deflection refers to the vertical displacement of structural elements under load. If floors, beams, or slabs deflect excessively:

- Internal partitions may crack or buckle
- Doors may jam or misalign
- Glass panels or cladding may fracture
- Ceiling systems may detach or deform
- Drainage slopes may reverse, leading to ponding or leaks

PBBD Design Approach:

- Engineers predict short-term and long-term deflection due to dead load, live load, and environmental effects.
- Load combination analysis ensures that total deflection remains within acceptable limits, often defined as:
    - L/250 for general floor spans (L = span length)
    - L/360 or stricter for brittle finishes like plaster or tiling
- In high-spec buildings (e.g., art galleries, laboratories), tighter limits may apply to prevent microcracking or movement-sensitive equipment misalignment.

Tools Used: Finite Element Analysis (FEA), 3D beam-slab interaction models, long-term creep simulation.

Vibration Sensitivity and Human Comfort

Modern structures are often lighter, longer-spanning, and more flexible than ever before. This increases sensitivity to vibrations caused by:

- Walking or dancing (in footbridges or event venues)
- Mechanical equipment (in hospitals or laboratories)
- Human movement in open-plan offices, gymnasiums, or auditoriums

Key Concepts:

- **Natural frequency:** Frequency at which a structure tends to vibrate; must not coincide with dominant loading frequencies (e.g., walking = 1.6–2.4 Hz).
- **Mode shapes:** The patterns of deformation during vibration; assessed to understand motion distribution.
- **Peak acceleration:** Measured in milli-g (mg), this determines perceived motion and comfort thresholds.

PBBD Mitigation Strategies:

- Tuned Mass Dampers (TMDs) or Viscous Dampers to absorb energy and reduce oscillation amplitude
- Structural tuning to shift the natural frequency out of sensitive bands
- Increased stiffness or mass (e.g., adding steel deck over concrete floor) to improve vibration performance

Comfort Criteria Standards:

- ISO 10137 and CIBSE Guide A for acceptable acceleration thresholds

- Office: ~15–20 mg
- Residential: ~5–10 mg
- Hospitals/labs: as low as ~2–5 mg

Example: A pedestrian bridge near a stadium was fitted with a tuned damper after tests showed footfall excitation near resonance, verified using full-scale testing and modelling in SAP2000.

Creep and Shrinkage

Creep and shrinkage are time-dependent deformations primarily associated with concrete, particularly in:

- Post-tensioned slabs and beams
- Tall vertical elements (e.g., walls and shear cores)
- Long-span roof structures

Definitions:

- **Creep:** Gradual deformation under sustained load
- **Shrinkage:** Volume reduction due to loss of moisture or hydration

These can result in:

- Long-term sagging
- Increased deflection beyond design expectations
- Loss of prestress force in tendons
- Cracking of finishes and rigid connections

PBBD Design Considerations:

- Use of time-step analysis tools (e.g., ETABS, ANSYS) to predict deformation over 10–50 years
- Application of Eurocode 2 creep and shrinkage models or more advanced time-dependent constitutive models
- Integration of construction sequencing and load history into structural modelling

Mitigation Tactics:

- Cambering (pre-curving) beams to counteract predicted sag
- Material selection (e.g., low-shrinkage concrete, additives)
- Strategic placement of movement joints

Thermal Expansion and Movement Accommodation

As structures are subjected to daily and seasonal temperature changes, thermal expansion and contraction must be allowed to occur without inducing stress that could:

- Crack facades or flooring
- Cause racking in steel frames
- Deform glazing or weather seals
- Distort mechanical systems or jointed slabs

PBBD Detailing Techniques:

- **Expansion joints:** Gaps between structural elements that allow for thermal movement, especially in long corridors, bridges, or facades.
- **Slip joints and flexible connections:** Allow movement at interfaces (e.g., between slabs and walls, steel and concrete connections)
- **Temperature gradient analysis:** Especially in sun-exposed roof decks or externally supported canopies

Tool Application: Thermal-structural analysis in 3D models, simulation of joint behaviour using ANSYS or Tekla Structures

Envelope Consideration: Thermal bowing of curtain walls is modelled to avoid cladding damage and ensure sealant flexibility.

PBBD Outcome: Holistic Structural Comfort and Usability

The goal of PBBD is to ensure that buildings feel and function well every day, not just during rare extreme events. Structural engineers ensure:

- Floor levels stay true and level
- Vibrations go unnoticed by occupants
- Partitions, ceilings, and finishes remain intact
- Long-term movement is predictable and harmless
- Environmental interaction (e.g., thermal effects) is absorbed without stress accumulation

This delivers not just compliance, but occupant satisfaction, building longevity, and reduced lifecycle maintenance costs.

Summary

Serviceability, comfort, and long-term performance are essential components of structural design in PBBD. Through:

- Predictive modelling of deflection, vibration, and creep
- Detailed movement and joint design and
- Occupant-focused performance thresholds

engineers deliver buildings that are not only strong but also refined, reliable, and user-centric—where technical excellence translates into everyday quality.

Resilience, Robustness, and Redundancy

In the era of climate change, global uncertainty, and increasing urban complexity, buildings must do more than resist collapse, they must endure disruption, protect occupants, and return to functionality with minimal downtime. In Performance-Based Building Design (PBBD), structural engineers take a proactive, data-driven approach to building performance continuity, combining resilience, robustness, and redundancy as pillars of safe and adaptable design.

These principles are especially critical in:

- Essential facilities (e.g., hospitals, data centres, fire stations)
- Infrastructure hubs (e.g., airports, bridges, rail terminals)
- High-occupancy buildings (e.g., stadiums, schools, towers)
- Post-disaster recovery zones where downtime is unacceptable

Defining the Three R's: Resilience, Robustness, Redundancy

- **Resilience:** The ability of a structure to absorb shock, recover functionality, and adapt to disruption.
- **Robustness:** The capacity to withstand abnormal conditions (accidents, misuse, overloading) without disproportionate failure.
- **Redundancy:** Providing multiple structural paths or systems so that failure of one does not compromise the whole.

These are not merely abstract goals, they are quantified, simulated, and embedded into the structural design process under PBBD.

Key Strategies for Resilience-Oriented Structural Design

Structural Redundancy – Load Path Diversity

Redundancy ensures that no single point of failure results in progressive collapse or catastrophic dysfunction. This is achieved by:

- Designing multiple vertical and horizontal load paths
- Creating alternate load transfer mechanisms between beams, slabs, and columns
- Avoiding over-reliance on one structural element (e.g., a single mega-column or cantilever)

Techniques:

- Continuous reinforcement in slabs and walls
- Secondary beams or backup frames to pick up redistributed loads
- Multi-core systems in high-rise towers for torsional stability
- Bridge-type frames that allow column removal without system failure

Assessment Tools: Disproportionate collapse analysis, alternate load path analysis, non-linear pushover simulations PBBD Insight: "If this critical column is lost, can the structure re-route the load safely?"

Energy Dissipation – Absorbing Shock and Preventing Collapse

Many disruptive events (e.g., earthquakes, blasts, heavy impacts) involve the rapid release of energy. Structural systems must dissipate this energy rather than transfer it into uncontrolled deformations or collapse.

Engineering Solutions:

- **Ductile Detailing:** Reinforced concrete and steel designed to yield and deform without brittle failure
- **Base Isolation Systems:** Bearings that decouple superstructures from ground shaking, protecting high-value or sensitive assets
- **Seismic Dampers:** Devices that convert vibrational energy into heat (e.g., viscous dampers, friction dampers, tuned mass dampers)
- **High-Damping Structural Materials:** Use of special alloys or composite systems in critical joints

Modelling Output: Reduction in floor accelerations, inter-storey drift, and stress concentrations under time-history loading.

Goal: Maintain structural and non-structural systems within safe deformation limits even under high-magnitude events.

Damage Containment – Designing for Predictable Failure

A resilient structure should fail gracefully, localising damage to sacrificial elements and preserving overall system integrity.

Design Features:

- **Structural Fuses:** Replaceable elements designed to deform or yield first (e.g., yielding links in eccentrically braced frames)
- **Breakaway Connections:** In curtain walls or cantilevered balconies that disengage under overload to protect the main structure
- **Segmented Components:** Designed to isolate cracking, absorb movement, or redirect force
- **Pre-planned Weak Links:** Areas engineered to absorb force and be replaced post-event (e.g., buckling-restrained braces)

Example: In bridges, shear keys or fuse bearings are engineered to break first, absorbing impact and preventing girder damage.

PBBD Question: "Can damage be isolated, and will it be easy to repair without deep structural intervention?"

Recovery-Driven Design – Rapid Return to Service

After a disaster, the speed of recovery can be as important as survival. PBBD enables structural engineers to consider repairability, replaceability, and modular recovery.

Principles:

- **Replaceable Non-Structural Components:**
  - Suspended ceilings, raised floors, and facades designed for rapid swap-out
  - Cladding designed with accessible fixings and modular panels
- **Service Core Protection:**
  - Structural isolation of vertical circulation shafts and critical services
  - Reinforced stairwells and fire-safe cores that remain operable post-event

- **Resilient Materials:**
  - Moisture- and fire-resistant insulation or finishes
  - Precast concrete units with damage-tolerant connections

Benefit: Reduced business interruption, lower insurance claims, and enhanced user trust in the built environment

Quantifying Resilience in PBBD

Engineers use simulation tools and performance frameworks to quantify resilience metrics such as:

Metric	Description
Functional Downtime	Time between event and full re-occupancy
Residual Drift	Remaining deformation after shaking or impact
Repair Cost Index	Percentage of building value requiring restoration
Service Retention Ratio	% of operational systems remaining functional post-event

These outputs help owners, insurers, and authorities make informed decisions about risk tolerance and investment in resilience measures.

Case Example: Resilient Mixed-Use Tower with Integrated Recovery Strategy

Location: Urban city centre with moderate seismic and wind exposure

Design Challenge: Maintain tenant operability within 72 hours post-disruption

PBBD Approach:

- Dual-frame structure with alternate load paths and outriggers
- Replaceable shear links and buckling-restrained braces at mid-height
- Dampers tuned to wind and seismic frequencies, reducing top-floor accelerations
- Prefabricated curtain wall designed for panelised replacement
- Modular rooftop MEP pods for easy replacement after lightning or storm impact

Outcome: The building met its design objective for Immediate Occupancy following a simulated 475-year seismic event and 100-year storm load.

Conclusion: The Resilience Edge of PBBD

PBBD empowers engineers to design buildings not only to survive disruptive events, but to continue functioning and recover swiftly through:

- Redundancy of systems
- Ductile, energy-absorbing mechanisms
- Targeted damage localisation
- Recovery-oriented detailing

Structures become adaptable, robust, and future proof, serving users continuously and dependably even under stress. Resilience is not just about building back better; it's about designing smart from the start.

Material Innovation and Lifecycle-Based Engineering

In traditional design, materials are selected primarily based on availability, strength, and cost, often adhering to conventional specifications and safety margins. In Performance-Based Building Design (PBBD), however, engineers are empowered to think beyond the prescriptive catalogue. They can specify non-traditional, hybrid, or sustainable materials provided they can demonstrate, through analysis, testing, and simulation that performance objectives are met or exceeded.

PBBD also introduces a lifecycle-based philosophy, encouraging engineers to consider not only initial strength and compliance but also durability, maintainability, environmental impact, and end-of-life adaptability. This approach aligns with broader global goals around net zero carbon, circular economy, and climate resilience.

Material Innovation in PBBD

Cross-Laminated Timber (CLT) and Engineered Wood Systems

CLT is a mass timber product made by bonding layers of lumber at perpendicular angles, offering high structural strength and dimensional stability.

PBBD Applications:

- Multi-storey constructions, including schools, apartments, and civic buildings
- Seismic design: Timber's lightweight nature reduces base shear, and energy dissipation can be designed via connection detailing

Performance Considerations:

- **Fire Resistance:** Demonstrated through charring rate calculations and full-scale furnace testing per EN 1995-1-2. The outer char layer insulates the inner core, preserving load-bearing capacity.
- **Dynamic Behaviour:** Seismic behaviour assessed using shake-table testing and hysteresis loop simulations
- **Acoustic and Thermal Benefits:** Naturally insulative and sound-absorbing, reducing energy use and improving occupant comfort

Engineering Toolsets: ETABS with timber modules, Dlubal RFEM, ABAQUS for non-linear wood modelling

Example: A six-storey CLT residential building in London was approved under PBBD by demonstrating 60-minute fire resistance and IO (Immediate Occupancy) seismic performance, despite being built from a combustible material.

High-Performance Concrete (HPC) and Fibre-Reinforced Composites

HPC and UHPC (Ultra-High-Performance Concrete) deliver increased strength, reduced permeability, and improved resistance to environmental degradation.

Advantages:

- Slim structural profiles (e.g., slender columns, thin floor plates), increasing usable space
- Extended service life, lower chloride penetration and freeze-thaw degradation
- Improved fire resistance and thermal mass for passive performance

Engineering Techniques:

- Tailored mix designs using silica fume, fibres (glass, steel, synthetic), and super-plasticisers
- Modelling of crack propagation, creep, and shrinkage under variable loading
- Dynamic load resistance for bridges, stadiums, and blast-resistant facilities

Tools: LS-DYNA for impact testing, COMSOL for thermal cracking simulation, and Abaqus for fracture analysis

Recycled and Low-Carbon Steel and Hybrid Systems

With embodied carbon increasingly under scrutiny, PBBD supports the selection of:

- Recycled steel, with lower environmental impact and similar strength characteristics
- Weathering steel (Corten) for long-term durability with minimal coating
- Steel-timber hybrids, combining timber's sustainability with steel's ductility and span capacity

PBBD Evaluation:

- Environmental Product Declarations (EPDs) and carbon footprint data used in material comparisons
- Structural performance modelled in tandem with One Click LCA or Tally to inform decision-making
- Thermal and corrosion performance assessed for coastal or humid environments

Impact: Use of low-carbon materials can reduce embodied CO_2 by 20–40% without compromising structural integrity when evaluated through lifecycle performance modelling.

Lifecycle-Based Engineering and Predictive Modelling

Durability Forecasts and Service Life Modelling

PBBD incorporates durability as a measurable design objective, not just a passive assumption.

Methods:

- Time-dependent deterioration models for corrosion, freeze-thaw, alkali-silica reaction (ASR), and sulphate attack
- Corrosion risk mapping using exposure classifications (per EN ISO 9223)
- Probabilistic deterioration analysis to assess failure likelihood over time

Output Metrics: Expected service life (years), time to first maintenance, cost of intervention, residual capacity over decades

Design for Deconstruction (DfD) and Adaptability

DfD is a core sustainability strategy under PBBD that focuses on how a structure can be disassembled, modified, or repurposed at the end of its primary use cycle.

Engineering Approaches:

- Bolted connections instead of welds for steel elements
- Dry-jointed precast concrete or modular floor systems
- Reversible fixings for timber assemblies
- Grid-based modularity to allow layout reconfiguration or extension

Applications: Office-to-residential conversions, post-disaster relocation of modular units, or material harvesting for reuse in future projects

Alignment: BS 8001 (Circular Economy Principles), ISO 20887 (Design for Adaptability and Disassembly)

Lifecycle Assessment (LCA) and Environmental Integration

Structural engineers under PBBD work alongside architects and sustainability consultants to balance structural performance with environmental impact across the entire lifecycle.

LCA Parameters:

- Embodied carbon (A1–A3 stages)
- Construction emissions (A4–A5)
- Use-phase performance (B1–B7)
- End-of-life recyclability (C1–C4)

Digital Tools: One Click LCA, eToolLCD, Athena Impact Estimator

Design Impact: Engineers may propose substituting 50% of cement with GGBS, designing for reuse of temporary structural steel, or using bamboo in compression-only truss elements, all justified through lifecycle modelling and simulation.

Example: Performance-Based Material Selection in an Educational Campus

Context: A university engineering building seeking low-carbon certification

PBBD Solution:

- Used a hybrid steel-CLT structure, reducing embodied carbon by 38% compared to a concrete frame
- Designed bolted, demountable connections to allow future wing expansion
- Evaluated all components via LCA software to guide procurement of low impact finishes and fasteners

- Modelled durability of below-ground reinforced elements for 100-year lifespan using chloride ingress simulations

Outcome: The project met BREEAM Outstanding and RIBA 2030 Challenge targets while exceeding structural performance requirements.

Conclusion: Material Intelligence in PBBD

Material selection in PBBD is performance-verified, lifecycle-conscious, and innovation-driven through:

- Engineering validation of non-traditional materials
- Predictive durability and deconstruction modelling
- Lifecycle assessment tools and circular design thinking

Structural engineers can make bold, informed choices that enhance performance, reduce environmental impact, and ensure long-term adaptability. Materials become not just structural components, but strategic assets in a resilient, sustainable, and future-ready building design framework.

Interdisciplinary Integration and Digital Modelling

In Performance-Based Building Design (PBBD), success is not defined solely by structural strength or code compliance, but by the synergistic performance of the entire building system. This requires interdisciplinary collaboration at every stage, architecture, structural engineering, fire safety, MEP, environmental systems, and even operations and maintenance.

Structural engineers in this environment no longer operate in isolation. Instead, they engage as digital collaborators, using advanced tools such as Building Information Modelling (BIM), parametric modelling, Finite Element Analysis (FEA), and Digital Twin technologies to deliver coordinated, intelligent, and responsive structural systems.

Why Interdisciplinary Integration Matters in PBBD

Traditional workflows often lead to fragmented systems, with late-stage clashes between architecture and engineering disciplines (e.g., duct runs through beams, inaccessible fire dampers, unaccounted-for structural depth). PBBD eliminates this by:

- Encouraging early and continuous coordination
- Validating performance objectives across all disciplines
- Enabling real-time feedback loops through shared models and simulations
- Supporting performance verification, not just assumption

Key Tools and Techniques for Digital Collaboration and Simulation

Finite Element Analysis (FEA)

FEA enables structural engineers to simulate and optimise complex interactions under real-world loading conditions. This is especially crucial in:

- Non-linear material behaviour (e.g., cracking concrete, yielding steel)
- Dynamic loads (e.g., seismic, wind, impact, thermal gradients)
- Complex geometries such as folded plates, space trusses, or shell structures

PBBD Use Cases:

- Foundation-soil interaction in tall or heavy buildings
- Long-span roofs subject to uplift, thermal movement, or snow accumulation
- Façade anchoring systems designed for lateral deformation, thermal bridging, or seismic racking

Tools: ANSYS, Abaqus, Robot Structural Analysis, SCIA Engineer

Outcome: Engineers don't just prove code compliance, they simulate actual behaviour and refine systems for efficiency, resilience, and long-term integrity.

Parametric and Generative Modelling

Parametric design allows engineers to:

- Define relationships and rules between geometry, loads, and performance
- Iterate rapidly across thousands of design options
- Optimise form and structure based on multi-objective criteria (e.g., strength, solar exposure, cost, aesthetics)

Tools:

Grasshopper for Rhino with plugins like Karamba3D (for structural simulation), Ladybug (for solar and energy modelling), or Millipede (for optimization).

Dynamo for Revit for rule-based model editing and coordination.

PBBD Application: A long-span roof with integrated solar fins was modelled using Grasshopper, adjusting structural section sizes based on solar exposure, wind uplift, and aesthetics in a single workflow.

Building Information Modelling (BIM) Coordination

In PBBD, BIM is not just a drawing platform, it's a live digital twin of the building that supports:

- Clash detection between structural framing, ductwork, pipework, fire zones, and lighting runs
- Design alignment between structural systems and non-structural elements (e.g., partition anchorage, slab edge conditions)
- Construction sequencing and phasing
- Data-rich objects that embed performance criteria (e.g., fire resistance, acoustic separation, thermal bridging, lifecycle info)

Coordination Tools:

- Autodesk Revit, Navisworks Manage, Solibri, Trimble Tekla BIMsight

Output: Structural engineers participate in coordination meetings (digital design charrettes), resolving spatial and performance conflicts before they reach the site, reducing RFIs, change orders, and construction risk.

Digital Twins and Real-Time Feedback Systems

A Digital Twin is a data-driven virtual model of a physical asset that evolves over time, informed by live sensors, usage data, and environmental feedback.

Applications in PBBD:

- Structural Health Monitoring (SHM): Embedding strain gauges, displacement sensors, and accelerometers into key elements
- Tracking deformation, cracking, corrosion, and fatigue in real time
- Predictive maintenance: Using real-world data to schedule inspections or trigger alarms
- Integration with AI-based diagnostic tools for anomaly detection

Example: A stadium roof includes embedded sensors feeding data into a cloud-based twin. Engineers monitor cable tension, snow load deformation, and wind response, updating FEA models and triggering alerts if thresholds are approached.

Interdisciplinary Workflows Enabled by PBBD Tools

Discipline	Coordination Focus	Digital Integration
Architecture	Form, aesthetics, daylight, layout	Rhino, Revit, BIM 360
MEP	Ducts, pipework, plant space	Navisworks, Revit MEP
Fire Engineering	Compartmentation, smoke modelling, fire resistance	CFD tools, BIM tagging
Structural	Load paths, deflection, vibration, anchorage	Revit, FEA, parametric tools
Façade / Envelope	Fixings, tolerance, movement, thermal bridging	3D detailers, performance simulators

Case Example: Parametric Truss and Solar Optimization in a Cultural Pavilion

Context: A lightweight cultural pavilion in a subtropical city required a long-span canopy that also shaded visitors and supported solar PV panels.

PBBD Strategy:

- Engineers used Grasshopper with Karamba3D to create a truss system responsive to:
    - Solar angle (for PV orientation)
    - Wind uplift (to minimise steel tonnage)
    - Structural deflection and vibration control
- Façade designers provided solar irradiance maps; structural sections were adjusted based on both load paths and thermal gain zones.
- Model exported to Revit for MEP coordination and clash resolution

Outcome: Delivered a 22% reduction in steel mass, improved PV efficiency by 14%, and met architectural goals for lightness and elegance.

Conclusion: Engineering Integration in the Digital Age

Interdisciplinary integration and digital modelling in PBBD are not just about aesthetics or automation, they are about:

- De-risking complexity
- Accelerating optimisation

- Improving cross-disciplinary performance
- Enabling post-occupancy adaptability through data

Through FEA, BIM, parametric workflows, and digital twins, structural engineers become digital collaborators, bridging conceptual vision with real-world behaviour and system alignment. The result is a smarter, safer, and more integrated built environment.

Conclusion: Structural Engineering in PBBD

In the context of Performance-Based Building Design (PBBD), structural engineering is redefined. Structural engineers are no longer viewed solely as guardians of structural stability or as technicians tasked with resisting gravitational and lateral forces. Instead, they take on a far broader and more influential role, designers of resilience, enablers of innovation, and stewards of sustainability.

They work at the intersection of engineering science, digital design, and performance simulation, ensuring that structures are not only compliant with minimum safety standards, but also optimised, adaptable, and future ready.

From Code-Checkers to Performance Strategists

In prescriptive frameworks, structural design has traditionally been constrained by fixed parameters and standardised loading conditions. PBBD empowers structural engineers to move beyond this, asking more meaningful questions:

- How will this structure behave during a rare seismic or wind event?
- What are the likely failure modes under overload, fatigue, or corrosion?
- How will occupants experience the building under normal use—vibration, sway, noise?
- Can we reduce embodied carbon while maintaining integrity?
- What happens to this structure in 10, 30, or 100 years?

By answering these questions through simulation, scenario modelling, and risk analysis, structural engineers become central to defining a building's long-term performance and lifecycle value.

Core Contributions of Structural Engineers in PBBD

Delivering Optimised, Project-Specific Solutions

Every building has a unique risk profile shaped by its geography, function, occupancy, and form. PBBD enables structural engineers to:

- Model site-specific seismic, wind, and soil conditions
- Customise structural systems (e.g., moment-resisting frames, damped outriggers, diagrids) to suit architectural intent
- Eliminate over-design and material waste by applying performance thresholds instead of blanket safety factors

Example: A coastal high-rise may integrate tuned mass dampers and aerodynamic shaping validated through wind tunnel testing—tailored precisely to its microclimatic wind pattern.

Enabling Material Efficiency Without Compromising Safety

With rising material costs and embodied carbon concerns, structural engineers must balance material efficiency with performance resilience. In PBBD, this is done through:

- Advanced materials (e.g., UHPC, CLT, recycled steel) with validated performance
- Finite Element Modelling (FEM) to optimise member sizing, reduce redundancy, and avoid over-engineering.
- Lifecycle-based decision-making: prioritising durability, reuse, and recyclability

Outcome: Leaner structures, reduced environmental impact, and improved design economy—without sacrificing safety margins.

Integrating with Architectural and Environmental Performance Objectives

PBBD dissolves the silos between disciplines. Structural engineers contribute directly to broader project goals, including:

- **Passive environmental strategies:** Coordinating structural form with solar orientation, wind paths, or natural ventilation
- **Architectural expression:** Enabling bold forms (e.g., cantilevers, curved shells, open atria) through advanced modelling and material innovation
- **Fire safety integration:** Designing fire-resistant load paths, protected egress zones, and structural robustness under thermal loading

Collaborative Design Tools: Shared BIM models, parametric scripting, and integrated performance dashboards ensure seamless cross-disciplinary alignment.

Future-proofing for Climate Extremes, Use Changes, and Adaptability

Structural engineers in PBBD anticipate not only first use but future use. They embed resilience and adaptability into design decisions by:

- Designing for extreme weather events (e.g., flooding, wind storms, heatwaves) using climate projections and load extrapolation
- Allowing for vertical or horizontal expansion, reconfiguration, or conversion (e.g., office to residential)
- Supporting design for deconstruction (DfD) through modular grids, reversible fixings, and non-welded joints

Lifecycle Thinking: Structural systems are evaluated not only for strength but for inspection intervals, repairability, retrofit compatibility, and ease of monitoring.

The Structural Engineer's Expanded Identity in PBBD

In PBBD, the structural engineer evolves into a multidisciplinary enabler who wears many hats:

Role	Description
Predictive Modeller	Simulates complex performance scenarios, from seismic time histories to long-term creep and fatigue
Risk Strategist	Designs for resilience, damage limitation, and business continuity under uncertain events
Design Integrator	Collaborates across architecture, fire, MEP, and sustainability to unify systems and avoid conflict
Sustainability Steward	Quantifies and reduces material impact, maximises structural lifespan, and enables reuse or repurpose
Digital Innovator	Leverages tools like FEA, BIM, digital twins, and generative algorithms to optimise and communicate design intent

Conclusion: Structural Design as a Driver of Performance, not a Constraint

In PBBD, the structural engineers are not the end-of-line verifiers of safety, they are core contributors to the building's intelligence, performance, and resilience by

- Modelling behaviour across timescales, intensities, and uncertainties
- Collaborating across disciplines and data platforms

- Prioritising comfort, adaptability, and efficiency

Structural engineers ensure that buildings don't merely stand, they endure shocks, accommodate change, sustain performance, and enhance human experience.

They are, ultimately, the quiet architects of structural intelligence, bringing vision to life through physics, simulation, and foresight.

Chapter 6

Mechanical and Electrical (M&E) Engineers in Performance-Based Building Design

In the realm of Performance-Based Building Design (PBBD), M&E engineers are no longer viewed solely as service specialists responsible for supplying power or conditioned air. Instead, they are key performance contributors, designing building systems that are not only efficient and functional but adaptable, resilient, and responsive to environmental and operational demands.

Their design decisions directly impact a building's:

- Energy performance
- Indoor environmental quality (IEE)
- Operational resilience
- Occupant well-being
- Carbon footprint and regulatory compliance

In PBBD, M&E systems are evaluated not just on specification but on measurable outcomes, from energy use intensity (EUI) and comfort metrics to recovery response time and fault tolerance.

Core Functions of M&E Engineers in PBBD

Focus Area: Heating, Ventilation, and Air Conditioning (HVAC)

HVAC systems are among the most energy-intensive and operationally significant components in any building. In Performance-Based Building Design (PBBD), the role of M&E engineers in HVAC design shifts from simply meeting static regulations to dynamically modelling, simulating, and optimising performance outcomes across a range of operating conditions, use scenarios, and climatic zones.

These engineers are responsible not just for indoor comfort, but also for energy efficiency, air quality, system robustness, and integration with the wider building performance objectives, including occupant satisfaction, carbon footprint reduction, and adaptive operation over the building's lifecycle.

Key Focus Areas in HVAC Design under PBBD

Thermal Comfort Modelling and Simulation

Instead of relying solely on standard temperature targets or air change rates, PBBD-driven HVAC design employs predictive comfort modelling to tailor indoor conditions to both human physiology and actual usage patterns.

Techniques and Parameters:

- **Predicted Mean Vote (PMV)** and Predicted Percentage Dissatisfied (PPD): Evaluate how occupants are likely to feel based on temperature, humidity, air speed, activity level, and clothing
- **Operative Temperature:** Combines air and radiant temperatures to reflect true perceived comfort
- **Localised Airflow Modelling:** Simulates ventilation effectiveness in specific zones (e.g., corner offices, atria, classrooms)
- **Tools:**
    - IES Virtual Environment (IES VE), DesignBuilder, EnergyPlus, and CFD plugins (e.g., FloVENT, SimScale) are used to assess:
        - Spatial temperature gradients
        - Stagnant air zones
        - Radiant asymmetry
        - Overheating risks by orientation or occupancy load

Outcome: HVAC systems are tuned for adaptive comfort, seasonal changes, and occupant expectations, rather than relying on static comfort bands alone.

Demand-Based Ventilation (DBV) and Adaptive Flow Control

Traditional systems often ventilate based on design occupancy or floor area. PBBD enables engineers to apply real-time demand-responsive ventilation, significantly reducing energy waste while ensuring air quality.

Features:

- CO_2 sensors detect occupancy levels and trigger increased fresh air intake
- Volatile Organic Compounds (VOC) sensors monitor pollutants from furniture, materials, or equipment
- Humidity sensors regulate flow rates in moisture-sensitive environments (e.g., museums, libraries, operating rooms)

Benefits:

- Reduces energy use by avoiding over-ventilation during partial occupancy
- Enhances indoor air quality (IAQ) by responding to actual contaminant levels
- Supports zoning and microclimate management for flexible workspaces or breakout areas

PBBD Objective: Align ventilation with real-world demand for precision air quality control and energy optimisation.

Energy Recovery Ventilation (ERV) and Heat Exchange Strategies

In PBBD, HVAC systems are expected not only to condition air but also to recover energy from exhaust air streams to reduce load on heating and cooling plants.

Key Technologies:

- Heat Recovery Wheels (enthalpy or sensible-only)
- Run-around coils and plate heat exchangers
- Desiccant-based systems for humidity-sensitive environments

These systems allow up to 75–85% thermal energy recovery, improving:

- Seasonal energy efficiency
- Part-load performance during shoulder months
- Carbon emissions associated with heating/cooling demand

Impact: Lower HVAC energy use per m^2, reduced carbon emissions, and improved cost efficiency in long-term operation

Resilience-Oriented HVAC Design

PBBD requires buildings to maintain a minimum level of functionality under abnormal conditions such as equipment failure, climate extremes, or fire events. HVAC systems must therefore incorporate resilience features including:

System Redundancy:

- N+1 chillers, boilers, or air handling units (AHUs) in mission-critical zones (e.g., healthcare, data centres)
- Cross-zonal ducting for emergency airflow redistribution
- Automated switchover controls via Building Management Systems (BMS)

Fire and Smoke Response:

- Smoke control dampers, stair pressurisation, and purge ventilation
- HVAC integration with fire detection systems to control air movement during fire scenarios

Post-Disaster Functionality:

- Ability to provide minimum ventilation or emergency cooling after power outage (e.g., via generator-supported fans)

Design Alignment: NHS HTM 03-01 (healthcare HVAC), BS EN 12101 (smoke control), and ISO 22301 (business continuity)

Performance Metrics and Validation

In PBBD, performance is not inferred, it is quantified, modelled, and verified. Key HVAC performance indicators include:

Metric	Description
Heating/Cooling Load (kW)	The capacity required to maintain thermal comfort at design conditions
HVAC Energy Use (kWh/m²/year)	Annual energy intensity for heating, cooling, ventilation, and distribution
Air Change Effectiveness (ACE)	A measure of how efficiently supply air displaces room air, often used in healthcare and cleanrooms
Temperature Stratification (°C/m)	Variation in temperature over height; ideally <1.5°C/m for comfort
System Coefficient of Performance (COP/EER)	Cooling or heating output per unit of energy input
Thermal Zoning Responsiveness	Ability of systems to maintain distinct conditions in separate building zones

Case Application: PBBD HVAC in a Mixed-Use High-Rise

Context: A 35-storey tower includes residential units, offices, and retail zones with mixed operating hours and comfort requirements.

PBBD HVAC Strategy:

- Simulated thermal comfort for each use type using DesignBuilder and ASHRAE 55 parameters
- Implemented demand-control ventilation based on CO_2, with override for high-load events (e.g., conferences)
- Used ERV systems in offices and retail to precondition fresh air, reducing chiller size by 18%
- Zoned AHUs per occupancy pattern with smart thermostats and occupancy-linked airflow
- Introduced resilience features including backup fan coils for common areas and automatic smoke purge in stairwells

Outcome: Building achieved thermal comfort compliance 95% of occupied hours, with HVAC energy intensity reduced by 27% compared to baseline models.

Summary: HVAC Design Excellence in PBBD

Through performance-based methods, M&E engineers transform HVAC systems into adaptive, efficient, and occupant-responsive infrastructures. Key takeaways include:

- Systems are simulated and verified for comfort, IAQ, and energy demand, not just code compliance
- HVAC components are coordinated with architectural design, occupancy patterns, and climate variability
- Smart controls, thermal zoning, and energy recovery become standard elements, not premium features
- Designs are validated using digital models before construction, and monitored post-occupancy to close the performance gap

Lighting and Daylighting Systems

In Performance-Based Building Design (PBBD), lighting design is more than simply specifying luminaires to meet target lux levels. It is a dynamic integration of natural daylight and artificial lighting, calibrated to deliver energy efficiency, visual comfort, occupant health, and operational resilience.

M&E engineers, working alongside architects and sustainability consultants, develop lighting strategies that are simulated, optimised, and validated against diverse performance metrics, ranging from glare control and light distribution to circadian rhythm support and emergency preparedness.

Their contributions are particularly critical in:

- Net zero carbon buildings
- Workplaces, healthcare, and learning environments
- Public spaces, galleries, and performance venues
- Buildings with complex geometries or dense urban contexts

Key Roles of M&E Engineers in Lighting and Daylighting Systems under PBBD

Daylight Autonomy Modelling

Daylighting reduces energy demand and enhances psychological well-being, but only when it is properly distributed and glare-free. PBBD requires engineers to simulate natural light levels, optimise façade elements, and balance daylight penetration with solar control.

Tools and Methods:

- Radiance, Honeybee+Ladybug (Grasshopper), IES VE, DIVA, and ClimateStudio
- Simulate daylight availability (spatial Daylight Autonomy - sDA) and useful daylight illuminance (UDI) to determine how much of the occupied time and area meet target daylight levels (typically 300–3,000 lux)
- Annual Sunlight Exposure (ASE) and Daylight Glare Probability (DGP) used to assess risk of visual discomfort, particularly in highly glazed or south-facing spaces

Design Applications:

- Light shelf placement
- Window size and transmittance selection
- Internal surface reflectance adjustments
- Shading systems (manual, fixed, or dynamic)

PBBD Performance Metrics:

- sDA(300/50%) > 55% of occupied area is a typical target
- ASE(1000/250) < 10% of floor area to minimise glare

Adaptive Artificial Lighting Control

Artificial lighting must be efficient, flexible, and responsive. PBBD places strong emphasis on integrating smart controls to reduce energy usage and tailor lighting to user needs and occupancy patterns.

Control Strategies:

- **Daylight-linked dimming:** Sensors adjust electric light output in response to natural daylight levels
- **Occupancy sensing:** Lights turn off or dim when areas are unoccupied, using motion or infrared detection
- **Task tuning:** Lighting is calibrated to specific tasks (e.g., 300 lux for circulation areas, 500–750 lux for reading or lab work)
- **Time-based control:** Lighting profiles follow schedules that reflect operational hours or programmed routines

Smart Integration: Lighting systems are often linked to the Building Management System (BMS) or IoT platforms for real-time optimisation, remote control, and performance logging.

PBBD Target: Lighting energy consumption is typically limited to <5 W/m^2, or <2.5 W/m^2 in ultra-low-energy designs.

Tuneable White and Circadian Lighting Systems

Lighting can support human health and productivity by aligning artificial light with the body's circadian rhythm, especially in schools, hospitals, and 24/7 facilities.

Features of Circadian-Aware Lighting:

- Tuneable white LED fixtures that shift colour, temperature and intensity throughout the day:
    - Cool white (5000–6500K) in the morning to promote alertness
    - Warm white (2700–3000K) in the evening to support relaxation
- Dynamic lighting sequences designed using equivalent melanopic lux (EML) or circadian stimulus (CS) metrics
- Systems may be adjusted based on daylight availability, time of day, user settings, or occupancy schedules

Use Cases:

- **In schools:** Enhances cognitive performance and sleep cycles in students
- **In healthcare:** Supports recovery and reduces delirium in ICU and elderly patients
- **In offices:** Boosts concentration and reduces fatigue in open-plan environments

Measurement: Luminance uniformity, colour rendering index (CRI >90), and circadian lighting scores (target CS > 0.3 for alertness periods)

Emergency Lighting and Resilience Systems

Lighting design must also address life safety requirements, particularly in fire, power outage, or emergency egress situations.

PBBD Integration Includes:

- Central battery or distributed power systems with 1–3 hour runtime
- Escape route lighting for stairs, corridors, and exits (typically ≥1 lux minimum)
- Open area anti-panic lighting to prevent disorientation in large public spaces (typically ≥0.5 lux)
- High CRI emergency lighting in medical or critical areas to support operational tasks during failure scenarios

Coordination with Fire Engineering: Lighting systems are designed to activate with fire alarms, remain visible through smoke (e.g., via photoluminescent markers), and avoid electrical interference with emergency comms.

Resilience Focus: Dual wiring routes, failover circuits, and integration with generator-backed power systems enhance reliability.

Lighting Design Outputs in PBBD

Engineers deliver quantified performance outputs that align with functional and visual criteria, including:

Metric	Description
Lighting Power Density (LPD)	Watts per square metre of lighting installed—used to assess energy efficiency
Uniformity Ratios (U_0)	Ratio of minimum to average illuminance—important for visual comfort and safety
Daylight Glare Probability (DGP)	Probability of visual discomfort due to contrast or brightness from daylight
Maintained Illuminance (lux)	Minimum light level over time considering depreciation—tailored to space type
Colour Rendering Index (CRI)	Quality of light colour rendering, typically >80 for general spaces, >90 for healthcare and museums

These parameters are iteratively modelled using BIM-linked simulation tools and validated through performance documentation aligned with:

- EN 12464-1 (indoor lighting)

- SLL Lighting Guides
- ASHRAE 90.1
- LEED and BREEAM daylighting credits

Example: Performance-Based Lighting Design in a University Library

Context: A new academic library required daylight-rich reading areas with energy efficiency and visual comfort across variable usage patterns.

PBBD Lighting Strategy:

- Simulated daylight performance using ClimateStudio and Radiance, optimising skylight layout and solar control louvers
- Integrated tuneable white lighting system that mimicked natural light cycles to support circadian health
- Used daylight-linked dimming across stacks and reading zones with occupancy sensors for low-traffic hours
- Maintained lux levels: 500 lux in reading zones, 300 lux in circulation areas, 750 lux in task rooms
- LPD < 4 W/m², with emergency lighting integrated into suspended linear fittings

Outcome: Achieved visual comfort compliance per EN 12464-1, received BREEAM Excellent daylight credit, and reported a 20% lighting energy reduction in post-occupancy analysis.

Conclusion: Lighting as a Performance System in PBBD

In PBBD, lighting is no longer just about visibility, it's about:

- Optimising user experience
- Supporting health and productivity
- Reducing operational energy
- Responding to real-world conditions dynamically

M&E engineers play a central role in modelling, validating, and implementing lighting systems that are intelligent, integrated, and human-centric, aligning architectural design with measurable environmental and user performance.

Electrical Power Distribution and Resilience

In Performance-Based Building Design (PBBD), electrical engineering extends far beyond the traditional provision of power to outlets and lights. Instead, it encompasses a strategic, predictive, and performance-driven approach that ensures systems are resilient, efficient, intelligent, and fail-safe, even under abnormal conditions.

Electrical power infrastructure must support:

- Uninterrupted functionality of critical systems
- Energy efficiency and operational optimisation
- Integration with on-site generation and storage
- Real-time diagnostics, load management, and automation
- Rapid recovery following fault or disaster events

This makes the role of electrical engineers in PBBD vital not just for code compliance, but for enabling performance continuity, smart energy use, and future-proof building operation.

Engineering Strategies for Power Distribution in PBBD

Load Profiling and Smart Power Distribution

Conventional designs often apply static, overestimated loads that result in inefficient sizing and energy waste. PBBD electrical design begins with dynamic load profiling, based on:

- Real-time energy use simulations
- Diversity factors for occupancy, equipment, and systems
- Peak vs off-peak consumption patterns
- Anticipated growth and plug load expansion

Smart Distribution Strategies:

- Zoned power circuits based on use type, load behaviour, or daylight availability
- Automated load shedding: Non-critical loads (e.g., signage, non-essential outlets) are disconnected during peak demand or emergencies
- Peak shaving and demand shifting: Through battery storage or flexible load activation, engineers flatten demand curves to avoid peak tariffs and reduce transformer stress

Design Tools: ETAP, PowerFactory, OpenDSS, and BIM-integrated simulation tools (e.g., IESVE, Revit MEP) PBBD Outcome: Avoids overdesign, reduces CAPEX on oversized switchgear and cables, and optimises energy use per occupant or use zone

On-Site Generation and Energy Storage Integration

PBBD encourages and often requires buildings to produce, store, and manage energy locally supporting energy independence and decarbonisation targets.

Typical On-Site Generation Systems:

- **Solar PV Arrays:** Integrated into rooftops, façades, or parking canopies
- **Combined Heat and Power (CHP):** For buildings with continuous thermal loads, such as hospitals or large mixed-use developments
- **Fuel Cells or Microturbines:** Low-emission, reliable power for critical infrastructure
- **Wind Turbines:** Where wind resources and site conditions permit

Energy Storage:

- **Battery Energy Storage Systems (BESS):** Smooth out solar intermittency, provide backup, and allow time-of-use tariff optimisation
- **Thermal energy storage:** Integrates with M&E systems to reduce electrical load spikes

Integration Considerations:

- Grid-tied vs off-grid functionality
- Battery discharge planning (e.g., for 4-hour vs 8-hour support)
- Microgrid management and seamless switching logic

Performance Metrics:

- Renewable contribution ratio
- Grid import reduction %
- System availability (uptime %)

Redundancy Planning and Failover Engineering

In mission-critical facilities (e.g., hospitals, data centres, airports), power system reliability is paramount. PBBD focuses on designing redundancy without inefficiency, ensuring that even under fault conditions, systems remain operational.

Key Redundancy Elements:

- **Dual electrical feeders** from separate substations or generators
- **Automatic Transfer Switches (ATS):** Seamless switching between power sources

- **Uninterruptible Power Supply (UPS):** Maintains power to critical systems during switchover or transient faults
- **Parallel Generator Systems:** For extended outages or grid failure

Design References: Uptime Institute Tier Classification (Tier I–IV), BS 7671 (IET Wiring Regulations), NFPA 70 and 110 (emergency systems)

PBBD Focus: Engineers analyse single points of failure, validate cascading failure prevention, and model restoration timeframes.

Typical Targets:

- Tier II (redundant capacity)
- Tier III (concurrent maintainability)
- Tier IV (fault-tolerant infrastructure)

Monitoring, Automation, and Predictive Diagnostics

PBBD integrates real-time intelligence into the electrical distribution system using:

Smart Monitoring Platforms:

- SCADA (Supervisory Control and Data Acquisition) systems for macro-level infrastructure control
- Building Management Systems (BMS) for subsystem monitoring (lighting, HVAC, plug loads)
- IoT-based smart metering for individual tenant, floor, or system analysis

Automation & Predictive Maintenance:

- Continuous condition monitoring of cables, switchgear, UPS, and transformers
- Alerts for voltage drops, harmonics, power factor drift, and thermal rise
- AI or rule-based algorithms that predict equipment failure or optimise energy scheduling

Data Visualisation: Dashboards display:

- Energy consumption by zone
- Load balance across phases
- Alerts on deviation from benchmarks
- Environmental impact (e.g., kg CO_2 saved per day)

Key PBBD Performance Metrics for Electrical Power

Metric	Description	Application
Power Usage Effectiveness (PUE)	Ratio of total facility energy to IT or mission-critical equipment energy	Data centres, labs
Energy Use Intensity (EUI)	kWh/m²/year consumed by electrical systems	Regulatory benchmarking (e.g., LEED, NABERS)
Electrical System Availability	% uptime over a 12-month period	Healthcare, airports
Fault Tolerance Level	Ability of system to continue operation under single or multiple fault conditions	Tier III/IV design evaluation
Downtime Impact Analysis	Financial or safety implications of electrical failure by system or zone	Business continuity planning

Case Example: PBBD Electrical Strategy for a Hospital Expansion

Context: A 12-storey hospital extension required continuous operation under all conditions, with zero tolerance for electrical interruption in surgical and ICU areas.

PBBD Strategy:

- Developed load profiles for 24/7 zones (ICUs, theatres), variable-use areas (offices), and non-critical zones (parking)
- Dual-feed switchboards with ATS to maintain continuity across feeder loss
- Separate UPS for life safety systems, critical IT servers, and BMS
- PV system with 300 kWh battery storage to reduce daytime peak demand
- Smart metering and SCADA dashboards tracking PUE, fault alarms, and energy mix (renewables vs grid)

Outcome:

- Exceeded NHS emergency power continuity requirements
- Achieved 99.997% power uptime
- Reduced annual grid electricity demand by 18% via on-site renewables

Conclusion: Electrical Engineering as a Pillar of Resilience in PBBD

In PBBD, electrical systems are designed for performance, continuity, and intelligence. Engineers go beyond layout and cable sizing to:

- Predict and model system resilience
- Integrate renewables and manage energy flows dynamically
- Prevent downtime with layered redundancy and smart diagnostics
- Support carbon reduction, smart grid readiness, and adaptive operation

They are not simply ensuring that power is "on," but that power is intelligent, reliable, efficient, and future-ready, enabling buildings to meet their performance targets under both normal and abnormal operating conditions.

Automation, Controls, and Smart Building Integration

In Performance-Based Building Design (PBBD), buildings are expected to respond, learn, and adapt. The static, fixed-output systems of traditional design are replaced with sensor-driven, real-time optimised systems that continuously balance comfort, energy performance, operational demands, and user preferences.

This paradigm shift is made possible by M&E engineers, who are now at the forefront of designing and integrating intelligent building systems. These include not only heating, cooling, lighting, and ventilation, but also fire safety, access control, security, and occupant-driven interfaces, all managed by sophisticated automation platforms.

Key Smart Systems in PBBD and the Role of M&E Engineers

Building Management Systems (BMS): The Operational Brain

A Building Management System (BMS) is the central nervous system of a modern building. It integrates and orchestrates multiple subsystems, including:

- HVAC (Heating, Ventilation, and Air Conditioning)
- Lighting and shading systems
- Power distribution and backup
- Security and access control
- Water systems
- Fire alarm and suppression

PBBD Integration:

- Real-time response to sensor data (e.g., temperature, CO_2, light levels, motion)
- Scenario-based control (e.g., energy-saving mode during low occupancy; enhanced ventilation during high CO_2)
- Time-of-use scheduling linked to daily operations, seasons, or event-based patterns
- Cross-system logic (e.g., lights dim when blinds open; HVAC off when windows are opened)

Smart Building Outcome: Optimised thermal comfort, lighting quality, and energy use, while maintaining safety and user satisfaction.

Digital Twins and Predictive Analytics

A digital twin is a live, data-driven virtual replica of a building that evolves with its operation. It leverages inputs from:

- IoT sensors
- BMS data streams
- User feedback platforms
- Environmental monitoring

M&E Contributions:

- **Linking physical systems to simulation models** for real-time comparison between expected and actual performance
- **Predictive maintenance:** Forecasting equipment failure based on trends in temperature rise, vibration, current draw, or usage hours
- **Energy drift detection:** Identifying deviations in energy use per system or zone compared to baseline models

Benefits in PBBD:

- Enables continuous commissioning
- Closes the gap between design intent and operational performance
- Provides a platform for adaptive control algorithms and self-learning systems

Example: An air-handling unit's fan motor begins showing vibration anomalies. The digital twin flags a likely failure within 30 days. Maintenance is scheduled, avoiding unplanned downtime.

Automated Fault Detection and Diagnostics (AFDD)

AFDD is a critical resilience layer in PBBD, ensuring that system faults are detected early, diagnosed correctly, and addressed promptly.

Applications:

- **HVAC systems:** Detecting stuck dampers, incorrect setpoints, sensor drift, or refrigerant loss
- **Lighting:** Noticing ballast degradation or zone control errors
- **Power systems:** Identifying phase imbalance, harmonic distortion, or generator testing failures

Engineering Features:

- Sensor redundancy and cross-checking logic to confirm fault legitimacy
- Prioritised alerting (e.g., critical faults override comfort alarms)
- Integration with maintenance management systems to automatically raise work orders

PBBD Impact: Ensures that systems deliver consistent, measurable performance, reducing energy waste and extending equipment life.

Zone-Based Environmental Controls and Personalisation

Performance-based design is about tuning environments not just for general use, but for specific, varied user needs across zones, floors, and time periods.

Features of Zone-Based Controls:

- **Individual temperature, lighting, and ventilation control** by space type or user group (e.g., conference room vs private office vs circulation corridor)
- **Occupancy-driven HVAC control:** Air volume, conditioning, and filtration vary based on live headcount or space booking data
- **Dynamic lighting zones:** Adjust light levels and colour, temperature based on daylight ingress, task type, or time of day
- **Acoustic zoning:** Smart control of partition dampers or white noise systems in sensitive areas (e.g., libraries, hospitals)

Design Tools: Zoning maps in BIM; simulation of zone response in tools like IES VE, EnergyPlus, or Honeybee; integration of user feedback platforms

Human-Centric Focus: Personal control enhances satisfaction, reduces complaints, and supports well-being, especially in hybrid workplaces or wellness-certified buildings.

PBBD Metrics Enhanced by Smart Integration

PBBD Goal	Smart System Impact	Key Metrics
Energy Efficiency	Adaptive lighting and HVAC scheduling	Energy Use Intensity (EUI), HVAC/lighting load (kWh/m^2)
Occupant Comfort	Zone-specific control, daylight tracking	PMV, lux levels, DGP, CO_2 ppm
Usability and Experience	BMS dashboards and occupant interface apps	User satisfaction, feedback loop scores
System Resilience	Redundancy and AFDD	System uptime %, mean time to failure (MTTF)
Carbon Emissions	Renewables integration, demand response	$kgCO_2e/m^2$/year, net zero alignment

Example: Smart PBBD Integration in a High-Performance Office

Project Brief: A tech firm's 10-storey headquarters with net zero aspirations and WELL certification goals.

PBBD Smart Strategy:

- BMS links HVAC, lighting, shading, security, and lift controls across 120+ zones
- Digital twin connected to real-time occupancy sensors, weather feeds, and internal feedback app
- AFDD identifies HVAC valve position faults, flagging systems that consume excess energy
- Circadian lighting adjusted in open-plan offices and meeting rooms
- App-based room booking integrates with climate and light setpoint pre-conditioning

Outcome:

- 23% reduction in annual energy compared to baseline
- 98.7% uptime across critical environmental systems
- Positive post-occupancy user experience score increased by 31% after rollout

Conclusion: Turning Systems into Performance Assets

In PBBD, building automation is not an add-on, it is a performance enabler. M&E engineers use automation and control strategies to:

- Ensure systems operate only when, where, and how they are needed
- Empower users to customise and optimise their environments
- Provide building managers with real-time visibility and predictive capabilities
- Support regulatory compliance, sustainability, and resilience benchmarks

With smart design, buildings are no longer passive containers, they are intelligent, responsive environments that learn, adapt, and evolve.

Integration with Sustainability and Environmental Systems

Modern buildings are expected to do more with less; less energy, less carbon, less water, and less waste, while delivering more comfort, intelligence, and value. This dual imperative lies at the heart of Performance-Based Building Design (PBBD), where M&E engineers are no longer service providers alone, but, more importantly, sustainability strategists and environmental performance enablers.

By aligning mechanical and electrical system design with environmental goals, M&E engineers ensure that buildings:

- Minimise operational energy and carbon
- Optimise comfort through passive and active design synergy
- Respond adaptively to real-world usage
- Maintain environmental performance over decades of use

How M&E Engineers Enable Environmental Performance in PBBD

Supporting Certification Goals: Net Zero, BREEAM, LEED, WELL

M&E engineers play a foundational role in achieving third-party environmental certifications. These certifications rely heavily on quantifiable system performance:

Key Contributions:

- Energy simulation and reporting for LEED EA Credits and BREEAM Energy Category
- Indoor Environmental Quality (IEQ) metrics, including air quality, thermal comfort, and acoustic criteria under WELL v2

- Commissioning and measurement planning, often led by M&E teams to validate post-occupancy performance
- Metering and submetering infrastructure, essential for credits under LEED and BREEAM for energy monitoring and demand management

Certifiable Outcomes:

- LEED v4 Energy and Atmosphere credits (EA Prerequisite: Minimum Energy Performance)
- BREEAM energy use intensity (ENE01), thermal comfort (HEA04), and ventilation (HEA02)
- WELL Features related to HVAC zoning, humidity control, and filtration (e.g., WELL Air, Thermal, and Mind categories)

PBBD Lens: Certifications aren't checklists, they reflect verified performance. M&E engineers ensure systems not only promise efficiency but deliver it consistently.

Embodied Carbon Analysis of M&E Systems

While structural materials like concrete and steel often dominate embodied carbon calculations, M&E systems can account for up to 30–50% of a building's embodied carbon in complex buildings (e.g., hospitals, labs, high-rise).

PBBD Practices:

- Embodied Carbon Lifecycle Analysis (ECLA) using tools like One Click LCA, EC3, or eToolLCD
- Inclusion of:
- Ductwork and diffusers
- Chillers, air handling units (AHUs), pumps
- Electrical conduits, cable trays, switchgear, and UPS
- Comparison of duct material choices (e.g., galvanised steel vs fabric), refrigerants (GWP potential), and cable types
- Design for disassembly or reuse to improve end-of-life carbon profiles

Embodied Carbon Benchmarks:

- CIBSE TM65: Embodied Carbon of Building Services Products
- RICS Whole Life Carbon Assessment Standard

PBBD Insight: M&E engineers guide material and system choices not only by lifecycle cost but lifecycle impact, closing the carbon loop from factory to disposal.

Whole Life Energy and Environmental Modelling

Beyond upfront efficiency, PBBD emphasises performance across the entire building life cycle, including:

- Operational Energy (regulated and unregulated)
- Replacement Cycles for M&E equipment (e.g., fan coils every 15 years, chillers every 25 years)
- End-of-Life Considerations: dismantling, recyclability, embodied energy in replacements

Modelling Methods:

- **Dynamic Simulation Models (DSMs):** Reflect hourly thermal and electrical profiles over decades, calibrated to local climate data
- **Scenario Testing:** Future climate files (e.g., 2050 and 2080), occupancy shifts, tariff changes
- **Carbon Intensity Forecasting:** Modelling carbon reductions over time as grids decarbonise

Key Outputs:

- Energy Use Intensity (EUI) in kWh/m^2/year
- Whole Life Carbon (WLC) in kgCO_2e/m^2
- Maintenance and replacement lifecycle cost
- Operational resilience under climate stress scenarios

PBBD Goal: Balance short-term performance gains with long-term environmental resilience and lifecycle return.

Passive Design Collaboration: Active + Passive Synergy

One of the most powerful aspects of PBBD is the integration of passive architectural strategies with active M&E systems, creating a performance loop between fabric and services.

Collaborative Focus:

- **Thermal Mass:** Engineers adjust HVAC response times and setpoints to leverage exposed concrete or brick for temperature moderation
- **Stack Ventilation:** Working with architects to position inlets/outlets, and size fans only for boost periods or winter recovery

- **Solar Shading & Glazing Design:** Engineers model HVAC loads based on daylight performance, overhang dimensions, and solar control glass
- **Envelope Airtightness:** Impacts ventilation strategies, infiltration modelling, and fan sizing

Collaboration Example:

- Architects simulate passive gains through IES VE or Ladybug
- M&E team uses these outputs to resize HVAC, optimise duct zoning, and reduce installed system capacity—lowering capex and OPEX

PBBD Benefit: Seamless alignment of passive and active systems, driven by shared modelling and performance criteria.

Case Example: Integrated M&E Design for a Net Zero Secondary School

Project Context:

A new-build secondary school in a northern UK climate sought net zero operational energy, full-year comfort, and WELL-aligned indoor quality.

PBBD M&E Strategy:

- Dynamic thermal modelling informed sizing of a low-temperature underfloor radiant system using ground-source heat pumps
- Applied demand-controlled mechanical ventilation with energy recovery wheels and night-purge ventilation
- Integrated a large rooftop solar PV array (100 kW) with on-site Lithium battery storage and a smart power management system
- LED lighting with zoned daylight dimming, occupancy sensors, and tuneable white circadian support in classrooms
- BMS with predictive control algorithms, drawing on weather forecasts, occupancy trends, and user override data

Performance Outcomes:

- Achieved EUI of 45 kWh/m^2/year
- 85% of school's energy met via on-site renewables
- Maintained high thermal and visual comfort across seasons
- Performance validated via post-occupancy testing and digital twin tracking

Conclusion: M&E Engineers as Environmental Performance Integrators

In PBBD, M&E engineers become critical agents of sustainability. Their contributions define the environmental success of buildings across time and usage cycles. They:

- Embed sustainability principles into core systems
- Use predictive models and simulations to drive evidence-based design
- Integrate with architectural form and structural resilience
- Ensure system efficiency, carbon intelligence, and climate adaptability

They are no longer just engineers of air and power, they are engineers of environmental intent, carbon balance, and building longevity.

Conclusion: The Evolving Role of M&E Engineers in PBBD

The practice of mechanical and electrical (M&E) engineering has undergone a profound transformation in the context of Performance-Based Building Design (PBBD). No longer limited to the layout of ductwork, the specification of HVAC units, or the sizing of electrical panels, M&E engineers are now central to the strategic, predictive, and adaptive performance of buildings.

They are emerging as architects of environmental intelligence, responsible for aligning building systems with operational realities, user expectations, and global sustainability goals.

From Systems Installers to Performance Integrators

Traditionally, M&E engineering was viewed as a service-oriented discipline that followed after architectural and structural design decisions. But in PBBD, M&E engineers are embedded from the earliest concept stage, where their insights and simulations inform:

- Passive design strategies
- Zoning logic
- Envelope performance
- Lifecycle cost models
- Carbon intensity trajectories

Their role is no longer reactive but predictive, ensuring that systems are not just functional, but optimal, responsive, and adaptable over time.

Key Dimensions of the Evolving M&E Role

Systems Optimisation for Comfort, Energy, and Resilience

M&E engineers design systems to meet far more than minimum code requirements. They:

- Deliver thermal and acoustic comfort through intelligent HVAC zoning and responsive controls
- Reduce operational energy through demand-driven ventilation, low-energy lighting, and smart distribution networks
- Build resilience into electrical and mechanical systems by planning for redundancy, failover, and recovery scenarios

They balance user well-being with system efficiency, ensuring that comfort does not come at the expense of energy performance.

Simulation-Driven Design Validation

A cornerstone of PBBD is that performance must be simulated, measured, and validated, not assumed.

M&E engineers:

- Run dynamic thermal and energy models to test how systems perform across seasons and occupancy variations
- Use computational fluid dynamics (CFD) to model airflow, heat transfer, and smoke movement
- Apply lighting simulations (e.g., Radiance, DIALux) to evaluate daylighting, glare, and artificial lighting distribution
- Conduct scenario-based testing for worst-case conditions (e.g., blackout, heatwave, partial occupancy)

This simulation-led approach provides clients and design teams with confidence that system performance will hold up under both typical and extreme conditions.

Multidisciplinary Collaboration and Integration

M&E engineers in PBBD operate at the nexus of all disciplines. They actively collaborate with:

- Architects, to harmonise services with spatial design, passive strategies, and user experience
- Structural engineers, to integrate duct routes, risers, plant loads, and equipment anchorage into the structural framework

- Sustainability consultants, to align systems with net zero carbon, BREEAM, LEED, and WELL targets
- Contractors and manufacturers, to verify that system specifications translate into viable procurement, installation, and commissioning

This collaborative mindset ensures that M&E systems are not in conflict with the building but are integrated assets that enhance design coherence and operational success.

Enabling Intelligent, Responsive, and Future-Ready Buildings

M&E engineers are the principal enablers of smart building functionality, embedding systems with the capacity to:

- Monitor and adjust to occupant behaviour, internal loads, and weather conditions
- Collect performance data to support digital twin models and predictive maintenance
- Adjust schedules, airflow, or lighting to respond to real-time analytics and occupancy patterns
- Integrate with renewable generation, energy storage, and grid interaction platforms

In doing so, they make buildings more than places to occupy, they make them dynamic environments that learn, respond, and optimise over time.

Bridging Infrastructure and Experience

Perhaps the most significant evolution in their role is that M&E engineers are now the link between hard infrastructure and human experience.

They translate infrastructure into:

- Thermal comfort
- Visual clarity
- Clean air
- Acoustic serenity
- Functional reliability
- Resilience during crisis

Through their work, buildings are not just technically operational—they are experientially exceptional.

Final Reflection: The PBBD M&E Engineer as a Strategic Designer

In PBBD, the M&E engineer is:

Role	Contribution
Systems Strategist	Designs multi-layered MEP systems tailored to performance scenarios
Environmental Modeller	Uses digital tools to test, validate, and optimise outcomes
Sustainability Enabler	Supports net zero carbon, WELL, LEED, and lifecycle optimisation
Collaborative Integrator	Aligns systems across architecture, structure, and operations
Innovation Catalyst	Delivers buildings that are intelligent, future-ready, and adaptive

They are no longer confined to "plant rooms and risers", they are instrumental in delivering measurable performance, user well-being, operational resilience, and environmental integrity.

As buildings become more dynamic, data-driven, and accountable to performance goals, the M&E engineer becomes an indispensable figure in shaping the next generation of intelligent, high-performing built environments.

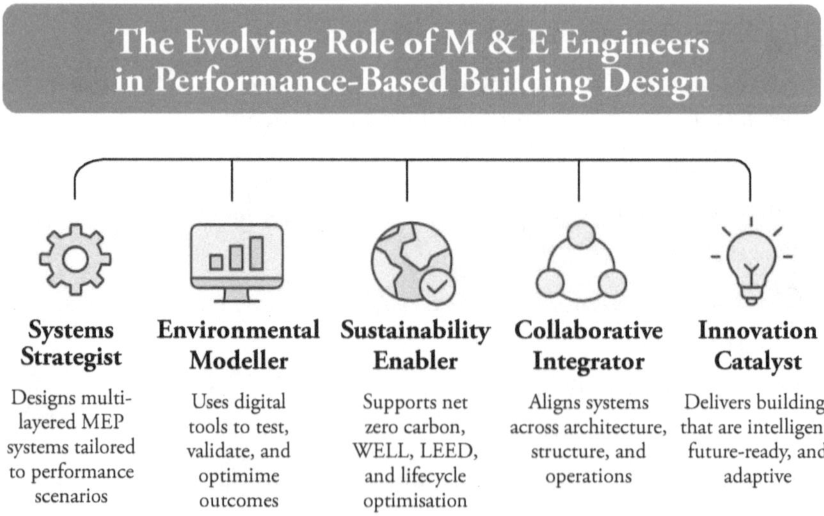

Chapter 7

The Role of Sustainability Consultants in Performance-Based Building Design

Sustainability Consultants are integral to the performance-based building design (PBBD) process, providing the expertise and analytical capabilities required to deliver buildings that meet environmental performance goals, occupant wellbeing standards, and long-term operational efficiency. Their input spans the full building lifecycle, from concept design through to post-occupancy evaluation, and ensures that sustainability is embedded not as an add-on, but as a core performance outcome.

By leveraging advanced modelling tools, aligning with internationally recognised standards (such as BREEAM, LEED, WELL, and the UK Net Zero Carbon Buildings Standard), and facilitating lifecycle assessments (LCA), Sustainability Consultants ensure that buildings achieve a high standard of environmental and social responsibility.

Environmental Modelling and Simulation

Environmental modelling and simulation lie at the heart of performance-based sustainability strategies. These advanced analyses allow design teams to predict and fine-tune how a building will perform under varying operational loads, occupancy patterns, seasonal changes, and climate scenarios, long before it is built. Sustainability Consultants lead these efforts, applying rigorous simulation tools to quantify the impact of design decisions on energy use, thermal comfort, daylight access, and indoor air quality.

In performance-based design, these models replace generalised compliance assumptions with detailed, building-specific data, enabling informed, adaptive, and resilient building outcomes. Below are key simulation domains where Sustainability Consultants deliver critical insights.

Dynamic Energy Modelling (DEM)

Dynamic Energy Modelling is a cornerstone of environmental simulation. It involves creating a virtual representation of the building, capturing geometry, orientation, material properties, occupancy schedules, lighting loads, equipment usage, and HVAC systems. Consultants use tools such as:

- IESVE (Integrated Environmental Solutions Virtual Environment)
- EnergyPlus (developed by the U.S. Department of Energy)
- DesignBuilder (a GUI-based platform that leverages EnergyPlus for simulation)

These tools calculate the building's hourly energy demand across a full year (8,760 hours), accounting for fluctuations in occupancy, internal gains, external weather, and system interactions.

Key Applications:

- Identifying peak energy loads, allowing HVAC systems to be sized precisely for demand rather than worst-case scenarios.
- Evaluating passive strategies, such as solar shading, natural ventilation, and thermal mass.
- Testing low- and zero-carbon technologies, like PV panels or ground source heat pumps, by modelling their interaction with building loads.
- Optimising building zoning, enabling temperature control based on use patterns (e.g., reducing HVAC loads in intermittently used areas).

Crucially, models are calibrated to local weather files (e.g., CIBSE TRY and DSY datasets for the UK) and can incorporate future climate weather projections to ensure buildings remain resilient under climate change scenarios.

Daylight and Solar Gain Analysis

Natural daylight is critical to both energy reduction and occupant wellbeing. Sustainability Consultants employ specialist tools such as Radiance, DIVA for Rhino, Dialux, and VELUX Daylight Visualizer to simulate how natural light penetrates interior spaces throughout the year.

Key Metrics Analysed:

- **Daylight Autonomy (DA):** Percentage of occupied hours a space receives adequate daylight.
- **Useful Daylight Illuminance (UDI):** Ensures light levels are not too low (underlit) or excessively bright (causing glare).
- **Spatial Daylight Autonomy (sDA):** Reflects how well daylight is distributed across a floor plan.
- **Annual Sunlight Exposure (ASE):** Measures the risk of visual discomfort due to over-illumination.

Applications in Design:

- Optimising window-to-wall ratios and glazing types to balance light penetration and heat gain.
- Designing shading devices, such as overhangs, fins, or electrochromic glazing, to mitigate solar glare and overheating.
- Daylight-linked controls, where artificial lighting dims or switches off based on available daylight, are modelled to forecast lighting energy savings.

These simulations directly support performance-based design by quantifying how design choices impact both energy performance and occupant comfort.

Overheating Risk Assessment

In response to climate change and urban heat island effects, overheating risk has become a major concern, particularly in highly insulated, airtight dwellings. Overheating is addressed through dynamic thermal simulations aligned with:

- **CIBSE TM59:** The UK's key guidance on overheating in homes.
- **CIBSE TM52:** Overheating criteria for non-domestic buildings.
- **Approved Document O:** A regulation introduced in the UK to mitigate overheating in new residential developments.

Sustainability Consultants simulate internal temperatures using DesignBuilder, IESVE, or TAS software. Inputs include:

- Building orientation and shading
- Thermal mass and insulation levels
- Glazing specifications and ventilation openings
- Occupant behaviour assumptions (e.g., window opening schedules, internal heat gains)

Outputs from these simulations determine:

- The number of hours per year a space exceeds thermal comfort thresholds (e.g., >26°C)
- Locations within the building most susceptible to overheating
- The effectiveness of passive cooling strategies, including natural ventilation, solar control, and night purging

Performance-based responses to these results might include adjusting façade geometry, introducing high-albedo roof materials, or re-zoning internal layouts to improve thermal conditions.

Natural Ventilation and Airflow Modelling

Natural ventilation modelling assesses the capacity of a building to maintain fresh air and remove excess heat without relying entirely on mechanical systems. This reduces energy use while improving indoor air quality and occupant wellbeing.

Consultants use:

- Computational Fluid Dynamics (CFD) for detailed 3D airflow analysis.
- Bulk airflow simulation tools, such as CONTAM, FloVENT, or Airpak, for macro-level studies.

Objectives include:

- Optimising cross-ventilation pathways through window sizing, placement, and alignment across spaces.
- Quantifying the stack effect and wind-driven ventilation in atria or stairwells.
- Testing hybrid ventilation strategies, combining natural and mechanical systems with automated controls.

This modelling supports decisions such as:

- Whether night-time cooling is sufficient to reset internal temperatures.
- Where to position louvres, vents, and trickle vents for effective airflow.
- How to avoid short-circuiting or "dead zones" where air stagnates.

Ventilation models also help evaluate compliance with Part F of the UK Building Regulations, ensuring that buildings meet minimum fresh air requirements while minimising energy penalties.

Thermal Bridging and Envelope Performance

Thermal bridging occurs where insulation continuity is compromised, typically at junctions, corners, window surrounds, and structural connections. These areas become weak spots in the envelope, leading to:

- Increased heat loss
- Condensation risks
- Reduced occupant comfort

Sustainability Consultants use 2D and 3D thermodynamic simulation tools, such as:

- THERM (developed by Lawrence Berkeley National Laboratory)
- Psi-Therm or Heat2

Simulation inputs include material properties, geometry, and boundary conditions. Consultants model:

- Linear thermal transmittance (Ψ-values)
- Surface temperature factors to assess condensation risk
- Overall envelope heat loss contribution from junctions (for SAP/SBEM compliance)

Through these models, they advise on:

- Thermal break detailing, e.g., in balconies or parapets
- Improved insulation continuity, such as wrapping insulation over structural elements
- Specifying advanced materials, such as vacuum insulation panels or aerogels where space is limited

Minimising thermal bridging supports both energy performance (lower U-values) and occupant health, particularly in airtight buildings where moisture control is critical.

Conclusion: A Data-Driven Foundation for Sustainable Design

Through rigorous environmental modelling and simulation, Sustainability Consultants enable design teams to make evidence-based decisions that improve building performance across energy, comfort, and resilience dimensions. These simulations form the basis of the performance-based approach, enabling tailored, predictive, and optimised strategies that reduce energy use, mitigate overheating, promote natural ventilation, and eliminate envelope weaknesses—ultimately driving sustainable, low-carbon, and occupant-centric building outcomes.

Lifecycle Assessment (LCA) and Embodied Carbon Modelling

Lifecycle Assessment (LCA) and embodied carbon modelling are critical tools in performance-based sustainable design. They enable the environmental performance of a building to be quantified across its full life cycle, informing material choices, construction methods, operational strategies, and end-of-life scenarios. Sustainability Consultants use these methods to guide clients toward low-impact, resource-efficient, and future-ready buildings, ensuring that environmental responsibility is embedded into design and procurement decisions.

In the context of performance-based design, LCA shifts the focus from prescriptive "tick-box" sustainability to measurable outcomes. These include reductions in embodied carbon, improvements in material durability and recyclability, and alignment with national and international carbon reduction frameworks, such as the UK Net Zero Carbon Buildings Standard, RICS Whole Life Carbon Assessment for the Built Environment, and the UKGBC Net Zero Roadmap.

Embodied Carbon Analysis

Embodied carbon refers to the total greenhouse gas (GHG) emissions associated with the extraction, processing, manufacture, transport, installation, maintenance, and disposal of building materials. Unlike operational carbon, which accrues over time, embodied carbon is largely locked in at the point of construction, making early design-stage decisions especially influential.

Sustainability Consultants use advanced LCA software platforms such as:

- One Click LCA
- eToolLCD
- Tally (Revit-integrated)
- Athena Impact Estimator
- GaBi or SimaPro (for custom industrial assessments)

These tools model the environmental impacts of the building's components using Building Information Modelling (BIM) or manual quantity take-offs, assigning carbon values (typically in $kgCO_2e$) to each material based on verified databases.

Applications:

- Comparing structural options (e.g., reinforced concrete vs. cross-laminated timber)
- Assessing façade systems, roofing materials, and internal finishes
- Identifying "hot spots" in the material palette that disproportionately contribute to the carbon footprint
- Prioritising high-impact interventions such as low-carbon concrete mixes, recycled steel, or bio-based insulation

Consultants often work closely with structural engineers and quantity surveyors to recommend low-carbon alternatives early in RIBA Stages 1–3, where design flexibility allows for significant carbon reductions at minimal cost.

Environmental Product Declarations (EPDs)

An Environmental Product Declaration (EPD) is a standardised, third-party verified document that quantifies the environmental impacts of a product based on a full LCA. These are developed in accordance with EN 15804 and ISO 14025, making them globally consistent and trustworthy.

Sustainability Consultants use EPDs to:

- Validate carbon claims made by suppliers and manufacturers
- Populate LCA models with accurate, product-specific data rather than generic assumptions
- Evaluate the performance of competing products within the same functional category
- Ensure materials align with Green Public Procurement, BREEAM, or LEED credits

Key EPD Indicators:

- Global Warming Potential (GWP) – i.e. embodied carbon
- Ozone Depletion Potential
- Acidification and Eutrophication
- Abiotic Depletion (non-renewable resources)
- Water Use and Waste Generation

Increasingly, UK-based clients and contractors are demanding EPDs as part of material submittals and some are developing internal databases of pre-approved low-carbon materials. Consultants support this process by maintaining up-to-date libraries and advising on the development of material procurement specifications that require or prioritise products with EPDs.

Whole Life Carbon (WLC) Reporting

Whole Life Carbon (WLC) encompasses both embodied and operational carbon. It captures the total emissions a building will generate from its inception to end-of-life, including construction, in-use emissions (heating, cooling, lighting, plug loads), maintenance, refurbishments, and demolition.

Sustainability Consultants prepare detailed WLC reports for clients, developers, and investors to:

- Demonstrate compliance with RICS guidance (2017)
- Align with the UK Net Zero Carbon Buildings Standard and UKGBC guidance
- Support planning applications and demonstrate alignment with local authority carbon targets
- Inform offsetting strategies, where net zero claims are made

WLC modelling typically covers the following stages as per EN 15978:

- **A1-A3:** Raw material supply, transport, and manufacturing

- **A4-A5:** Transport to site and construction
- **B1-B7:** Use phase, including energy, water, maintenance, and replacements
- **C1-C4:** End-of-life (demolition, recycling, landfill)
- **D (optional):** Benefits and loads beyond the system boundary (e.g. material reuse)

The WLC report includes:

- A visual breakdown of carbon by building component and life cycle stage
- Carbon intensity benchmarks (e.g., $kgCO_2e/m^2$ GIA)
- Sensitivity analyses based on future climate and refurbishment scenarios
- A roadmap for reducing emissions through design changes, procurement, and renewable energy sourcing

These reports help organisations benchmark their carbon intensity, communicate impact transparently, and meet ESG (Environmental, Social, and Governance) reporting requirements.

Circular Economy and End-of-Life Planning

The transition to a circular economy is fundamental to reducing embodied carbon and resource depletion in the built environment. Performance-based sustainability design supports this by designing out waste, maintaining material value, and prioritising reuse.

Sustainability Consultants support circularity by:

- Designing for disassembly: Encouraging bolted or dry-fixed connections over adhesives and welded joints
- Specifying modular systems that can be reconfigured or expanded without demolition
- Advising on "material passports" or digital twins that record material provenance, maintenance history, and disassembly instructions
- Assessing recycled content and reusability of key materials (e.g. raised access floors, partitioning systems, steel frames)
- Identifying reclamation opportunities, such as using salvaged bricks or timber from deconstruction projects
- At end-of-life, consultants model waste management routes and their carbon implications:
    - Recycling vs. landfill vs. incineration
    - Potential carbon savings from closed-loop recycling
    - Impacts of transport logistics and deconstruction techniques

This approach ensures that buildings contribute positively to resource efficiency, material recovery, and reduction of landfill dependency, supporting broader UK goals such as the Construction Resources and Waste Strategy and the Circular Economy Package under the Environment Act 2021.

Conclusion: Enabling Low-Carbon, Long-Life, Resource-Efficient Buildings

Lifecycle Assessment and embodied carbon modelling enable designers and clients to move beyond short-term compliance and embrace a holistic, data-driven view of sustainability. Through the work of Sustainability Consultants, performance-based design evolves from theoretical aspiration to measurable environmental stewardship. By quantifying and reducing emissions across every stage of the building lifecycle, these assessments underpin the transition to net zero, circular economy, and climate-resilient buildings, setting a new standard for the environmental performance of the built environment.

Certification and Standards Alignment

One of the core responsibilities of Sustainability Consultants in performance-based building design (PBBD) is the navigation, coordination, and successful delivery of environmental certifications and building performance standards. These frameworks provide third-party verification that a building meets, or exceeds, specific benchmarks for environmental quality, resource efficiency, occupant wellbeing, and climate impact.

Beyond simply "checking boxes," Sustainability Consultants ensure that certification is fully embedded into the design intent, construction process, and operational strategy—serving as both a compliance mechanism and a tool for design excellence. The integration of these certifications supports regulatory alignment, enhances asset value, reduces operational costs, and provides credible ESG credentials for owners and investors.

The following outlines the key certification schemes typically managed by Sustainability Consultants, along with their roles in facilitating performance-based compliance and environmental optimisation.

BREEAM (UK – Building Research Establishment Environmental Assessment Method)

BREEAM is the UK's most widely used and influential sustainability assessment method for buildings, with international variants and adaptations. It evaluates a building's performance across ten core categories, including energy, water, health and wellbeing, transport, materials, waste, land use and ecology, pollution, resilience, and innovation.

Key Consultant Responsibilities:

- **Pre-Assessment and Strategy Formation:** At RIBA Stage 1 or 2, consultants carry out a pre-assessment to establish the targeted BREEAM rating (Pass to Outstanding) and identify achievable credits. This becomes the project's sustainability roadmap.
- **Design Stage Submission:** Consultants coordinate with architects, engineers, planners, and contractors to ensure that credit evidence is embedded into specifications and drawings. They prepare the documentation for submission to BRE for interim certification.
- **Credit Management:** Using tools like the BREEAM Credit Tracker or project-specific matrices, consultants oversee the tracking, documentation, and verification of credits. These may include:
    - ENE 01 (Energy): Dynamic simulation modelling and Part L compliance.
    - HEA 04 (Thermal Comfort): Overheating analysis in line with TM59.
    - MAT 01 (LCA): Whole building lifecycle assessment with tools like One Click LCA.
    - MAN 03 (Responsible Construction): Contractor engagement and site monitoring.
- **Post-Construction Stage (PCS):** Consultants compile the as-built evidence, site photographs, commissioning reports, and contractor documentation required to complete the final certification process.

Integration with PBBD: BREEAM works synergistically with performance-based design by rewarding measurable outcomes rather than prescriptive checklists, especially in categories such as energy modelling, thermal comfort, daylighting, and LCA. It encourages early simulation, lifecycle thinking, and innovation, all core tenets of PBBD.

LEED (US/Global – Leadership in Energy and Environmental Design)

LEED is the globally recognised green building certification developed by the U.S. Green Building Council (USGBC). It evaluates sustainability across Location & Transportation, Sustainable Sites, Water Efficiency, Energy & Atmosphere, Materials & Resources, Indoor Environmental Quality, and Innovation.

Key Consultant Responsibilities:

- **Gap Analysis and Feasibility:** Consultants assess how the project aligns with LEED prerequisites and identify additional credits required to reach the desired rating (Certified to Platinum).
- **Credit Interpretation Requests (CIRs):** For novel strategies or ambiguous cases, consultants liaise with USGBC/GBCI to obtain rulings on acceptable compliance pathways.
- **ASHRAE Compliance:** For energy-related credits (e.g. EA Credit: Optimize Energy Performance), consultants ensure HVAC, lighting, and envelope designs align with ASHRAE 90.1 and perform energy modelling to show percentage energy savings.
- **Water Efficiency:** Consultants specify ultra-low flow fixtures, conduct usage calculations, and evaluate greywater reuse options to achieve points under indoor and outdoor water use reduction credits.
- **Materials Transparency:** LEED rewards supply chain transparency through EPDs, Health Product Declarations (HPDs), and Material Ingredient Reporting, areas where consultants ensure supplier documentation meets credit thresholds.

Integration with PBBD: LEED credits require quantified performance evidence in energy, water, IAQ, and material lifecycle impact. The consultant's role in LEED aligns seamlessly with PBBD, particularly in managing simulations, selecting verified products, and promoting passive/renewable strategies grounded in modelling.

WELL Building Standard (International WELL Building Institute)

The WELL Building Standard is a performance-based system that focuses on human health, wellbeing, and productivity. It evaluates a building's effect on physical, psychological, and emotional wellness through ten categories: Air, Water, Nourishment, Light, Movement, Thermal Comfort, Sound, Materials, Mind, and Community.

Key Consultant Responsibilities:

- **Preconditions vs. Optimisations:** Consultants identify the required preconditions and optional optimisation strategies needed to reach Silver, Gold, or Platinum WELL certification levels.
- **Sensor Deployment and Monitoring:** Consultants specify and oversee the installation of IAQ sensors, circadian lighting controls, acoustic treatments, and water filtration systems—all of which require real-world performance validation.

- **Policy Development:** WELL requires organisational policies for wellness programs, healthy food choices, biophilia, and mental health promotion. Consultants often support HR, workplace strategy, and building management teams in meeting these requirements.
- **WELL Performance Testing:** Unlike design-only certifications, WELL includes on-site post-occupancy testing (conducted by WELL Performance Testing Agents). Consultants prepare the site and facilitate coordination with the verification team.

Integration with PBBD: WELL is inherently performance-based—its metrics depend on measurable outcomes for occupant experience and comfort. It dovetails with PBBD strategies by prioritising dynamic thermal comfort modelling, daylight availability, ventilation effectiveness, and ongoing sensor-driven optimisation.

Net-Zero and Carbon Certification

Sustainability Consultants are also central to defining and verifying a building's net-zero status, whether net-zero energy, net-zero operational carbon, or net-zero whole life carbon.

Relevant frameworks and guidance include:

- UK Net Zero Carbon Buildings Standard
- UKGBC Net Zero Operational Carbon Framework
- CIBSE TM65 / TM54
- Passivhaus Plus / Premium (energy-focused net-zero approaches)
- Science-Based Targets Initiative (SBTi) for commercial portfolios

Key Consultant Responsibilities:

- **Carbon Boundary Definition:** Consultants define what is included in the carbon scope—regulated vs. unregulated loads, occupant emissions, mobility, refrigerant leaks, etc.
- **Modelling and Verification:** Consultants simulate annual operational energy demand and onsite/offsite renewable contributions using tools like IESVE or EnergyPlus. They validate performance with post-occupancy monitoring (POE) and metered data where possible.
- **Procurement Strategy:** Consultants verify the procurement of renewable energy through Power Purchase Agreements (PPAs), green tariffs, or Renewable Energy Guarantees of Origin (REGOs).

- **Carbon Offsetting:** For residual emissions, consultants source verified carbon offsets (e.g. Gold Standard, Verified Carbon Standard) and document the alignment with offsetting principles (e.g. permanence, additionality, transparency).

Integration with PBBD: Net-zero verification is inherently performance-driven, it requires full lifecycle thinking, integrated energy design, and robust measurement. Sustainability Consultants bridge the gap between design intent and real-world operation, delivering accountability through simulation, testing, and post-occupancy calibration.

Conclusion: Certification as a Tool for Performance Delivery

For performance-based building design, certification frameworks are not bureaucratic hurdles, they are strategic instruments for tracking, validating, and communicating performance. Sustainability Consultants ensure that these systems are fully integrated into the design workflow, supporting both compliance and innovation. By aligning projects with BREEAM, LEED, WELL, and Net-Zero frameworks, consultants ensure that sustainability outcomes are:

- Transparent and third-party verified
- Quantitative and measurable
- Relevant to climate, user, and regulatory contexts
- Aligned with global sustainability goals

This approach enhances project credibility, supports regulatory approval, and enables owners and developers to deliver buildings that are resilient, healthy, energy-efficient, and future-proofed.

Passive Design and Renewable Energy Integration

In performance-based sustainability design, energy conservation is pursued through a structured energy hierarchy: first reducing demand, then optimising systems, and finally generating low-carbon energy. At the foundation of this hierarchy lies passive design, a philosophy that minimises the need for mechanical intervention by leveraging the natural characteristics of climate, form, and material behaviour. Only after passive strategies have been fully exploited do Sustainability Consultants evaluate the integration of renewable and low-carbon technologies to meet the remaining energy demand.

This approach aligns with key UK and international frameworks *(e.g. The Energy Hierarchy by the UK Energy Institute, Part L of the Building Regulations, CIBSE TM52/*

TM59, UKGBC's Net Zero Carbon Framework) and represents a fundamental principle of performance-based building design.

Passive Solar Design

Passive solar design harnesses the sun's energy to reduce heating and cooling loads without active systems. Sustainability Consultants work with architects and building physicists to ensure that a building's form, orientation, and envelope design take full advantage of solar energy in winter while mitigating overheating risks in summer.

Key Passive Solar Techniques:

- **Orientation and Zoning:** Maximising southern exposure (in the northern hemisphere) to allow for winter sun penetration into living or working spaces, while locating service areas on less solar-exposed façades.
- **Thermal Mass:** Incorporating materials such as concrete, brick, or phase-change materials to absorb solar heat during the day and release it at night, flattening temperature swings and reducing heating demand.
- **Optimised Glazing Ratios:** Balancing solar gain, daylight availability, and heat loss by carefully calculating window-to-wall ratios by façade. This often involves increasing glazing on southern façades and reducing it on east/west to avoid low-angle sun glare.
- **Fixed and Dynamic Shading Devices:** Horizontal overhangs, vertical fins, brise-soleil, and dynamic systems (e.g. electrochromic glass, external blinds) are used to prevent overheating in summer while allowing low-angle sun in winter.

Simulation Tools:

- IESVE Radiance, Ladybug Tools for Rhino/Grasshopper, and EnergyPlus are used to simulate solar exposure, daylight penetration, and solar gain throughout the year, guiding iterative optimisation.

Envelope Optimisation

The building envelope is the most significant interface between the internal and external environment. A high-performance envelope reduces reliance on active systems by limiting unwanted heat transfer and air leakage, supporting thermal comfort, energy efficiency, and building durability.

Key Envelope Performance Strategies:

- **U-value Optimisation:** Consultants advise on material choices and wall/roof/floor assemblies to achieve or exceed regulatory targets. U-values are balanced with buildability, structural implications, and cost.
- **Thermal Bridging Mitigation:** Consultants work with structural and façade engineers to identify junctions prone to thermal bridging using 2D/3D simulation tools like THERM or Psi-Therm. Thermal breaks, insulated cavity closers, and continuity of insulation layers are specified to reduce cold bridging.
- **Airtightness Targeting:** Sustainability Consultants assist in setting airtightness targets (e.g., $\leq 3\ m^3/h{\bullet}m^2$ at 50 Pa for new builds) and recommend detailing practices and testing protocols to ensure compliance. Achieving high airtightness is particularly critical for low-energy buildings (e.g. Passivhaus).
- **Ventilation Strategy Integration:** Airtight envelopes must be balanced with sufficient ventilation. Consultants ensure compatibility between envelope detailing and natural/mechanical ventilation systems, avoiding condensation and ensuring healthy IAQ.

Envelope Design in PBBD: Performance-based envelope design involves dynamic thermal modelling, rather than relying on static U-value prescriptions. Consultants test how the envelope performs under real-time seasonal loads, using TM59 (overheating) and TM52 (thermal comfort) criteria to refine the passive design.

Renewable Energy Feasibility Studies

Once demand has been reduced through passive strategies, Sustainability Consultants evaluate the most appropriate renewable and low-carbon technologies for the site. This analysis is site-specific and performance driven, incorporating climate data, available space, local regulations, and energy profiles.

Technologies Evaluated:

- **Solar Photovoltaics (PV):** Roof- or façade-mounted systems sized based on irradiance, roof pitch/azimuth, structural load capacity, and shading. Consultants model annual yield (kWh/kWp), self-consumption ratios, and grid export implications.
- **Solar Thermal:** Assessed for domestic hot water (DHW) load offset, often in residential and leisure applications. Consultants evaluate tank sizing, orientation, and integration with backup systems.

- **Heat Pumps:** Ground-source (GSHP) and air-source heat pumps (ASHP) are modelled for seasonal performance factors (SPF), installation constraints, and local incentives (e.g., Boiler Upgrade Scheme in the UK). GSHP viability includes soil condition and borehole depth analysis.
- **Wind Turbines:** On-site wind energy is only feasible in specific rural or exposed urban locations. Consultants analyse wind speed frequency data and turbulence intensity to determine viability.
- **Biomass:** Typically limited to rural projects due to air quality and storage considerations. Consultants assess biomass against air pollution standards (e.g., Clean Air Zones).

Modelling Outputs:

- Energy Yield Predictions (e.g., PVsyst or RETScreen)
- Storage Scenarios (integration of battery systems or thermal storage)
- Grid Interaction Models (export limits, demand shifting)
- Return on Investment (ROI) and Payback Periods
- Carbon Savings per Technology, supporting Net Zero frameworks

Low-Carbon HVAC Strategy

Following demand reduction and renewables integration, mechanical systems are specified to complement the building's passive and active energy strategy. Sustainability Consultants collaborate closely with M&E engineers to select energy-efficient, low-carbon systems that are appropriately sized and responsive to actual loads.

Common Low-Carbon HVAC Approaches:

- **Variable Refrigerant Flow (VRF) Systems:** Allow for simultaneous heating and cooling across zones, with efficient heat recovery. Consultants assess zoning logic, refrigerant volume limitations, and efficiency in transitional seasons.
- **Heat Recovery Ventilation (HRV/MVHR):** Essential in airtight buildings. Consultants specify units based on airflow, heat exchange efficiency, and location of ducting within the building geometry. Filtration levels are also specified based on local air quality.
- **District Energy Networks:** Where available, consultants evaluate the carbon intensity, reliability, and regulatory status of district heat networks (e.g., whether they qualify as "low-carbon" under SAP calculations).

- **Zoning and Controls Integration:** Consultants recommend zoned heating/cooling and smart controls that integrate with BMS systems to enable demand-controlled ventilation (DCV), occupancy sensing, and dynamic setpoint adjustments.
- **Simulation and Optimisation:** All HVAC strategies are tested through dynamic simulation modelling to avoid over-specification (e.g., oversized cooling plant) and to validate real-time responses under variable occupancy and weather conditions.

Conclusion: Designing Buildings that Work with, Not Against, Nature

Sustainability Consultants champion a performance-based approach that begins with minimising demand through climate-responsive design, continues with optimised envelope and system design, and culminates in renewable energy integration tailored to the project's specific energy profile and location. This method aligns with the fundamental principle of "fabric first" design and ensures that buildings are not only low-energy in theory but verifiably high-performance in operation.

By strategically sequencing passive design and renewable integration, consultants help ensure that buildings are:

- Low-energy, low-emissions, and low-maintenance
- Comfortable and healthy for occupants
- Financially efficient over the life cycle
- Compliant with Net Zero Carbon and regulatory standards

Interdisciplinary Collaboration and Decision-Making

In performance-based building design (PBBD), sustainability outcomes are never the result of a single discipline or decision, they are the product of ongoing, coordinated, interdisciplinary collaboration. The complexity of modern building systems, regulatory frameworks, environmental targets, and stakeholder expectations means that environmental performance must be embedded across all aspects of the project lifecycle, from concept to post-occupancy.

Sustainability Consultants play a central role in this integrated process. Acting as facilitators, analysts, and translators, they work at the intersection of architecture, engineering, planning, and construction, ensuring that every key decision supports the project's environmental goals, whether those relate to energy, carbon, health, biodiversity, resilience, or regulatory compliance.

Their expertise allows them to ensure that sustainability is not treated as an "add-on," but as a core performance outcome that influences, and is influenced by, every other design decision.

Integrated Design Reviews and Multidisciplinary Workshops

Integrated design reviews are a cornerstone of collaborative sustainability. These are structured workshops or coordination meetings where representatives from all disciplines, architecture, structural, M&E, fire, cost, planning, acoustics, and beyond, convene regularly to review the design against performance targets.

Sustainability Consultants lead or contribute to these sessions by:

- **Setting environmental priorities early:** e.g. targeting a specific BREEAM rating, net-zero carbon, or WELL certification.
- **Bringing performance data to the table:** Using outputs from energy models, daylight analysis, or LCA tools to inform real-time decisions.
- **Challenging assumptions:** If a design choice increases solar gain or embodied carbon, the consultant ensures this is recognised and addressed.
- **Ensuring spatial and system coordination:** For example, confirming that ventilation ducts, shading devices, or PV panels can be physically accommodated without conflict.

Benefits of Integrated Reviews:

- Promote early-stage alignment between disciplines
- Avoid design silos that can lead to expensive retrofitting or rework
- Foster a shared understanding of sustainability goals across the team
- Facilitate faster, better-informed decision-making through evidence-based inputs

Design reviews are particularly powerful when aligned with RIBA Plan of Work stages, allowing Sustainability Consultants to "lock in" key strategies (e.g. orientation, MEP zoning, envelope detailing) before the window of influence narrows.

Value Engineering for Sustainability

Value engineering (VE) is often perceived as a cost-cutting exercise, but in a performance-based sustainability context, VE is a tool for balancing cost, performance, and long-term value. Sustainability Consultants work alongside cost consultants and the design team to ensure that VE decisions do not compromise critical performance outcomes, and instead, optimise them.

Key contributions to value engineering include:

- Cost–performance analysis of systems or materials, comparing options based on lifecycle cost, carbon savings, durability, and operational efficiency—not just capital cost.
- Carbon-informed VE, where materials are evaluated for their environmental impact as well as cost (e.g. low-carbon concrete mixes vs. standard mixes, or timber vs. steel frame systems).
- Prioritising "performance per £": Advising clients where sustainability spend is most effective, e.g. enhanced insulation vs. premium finishes, or battery storage vs. extra PV capacity.
- Supporting grant or incentive alignment: Identifying where low-carbon systems might qualify for government support (e.g. Boiler Upgrade Scheme, SALIX funding, or local authority green funds).
- Rather than removing sustainability features to save costs, consultants use VE to:
    - Optimise return on investment (ROI) in energy/carbon terms
    - Phase investments over time (e.g. design for future renewables even if not installed immediately)
    - Substitute lower-impact solutions that still meet design intent (e.g. switching cladding materials with lower embodied carbon)

This approach reinforces the performance-based mindset, emphasising outcomes rather than default specifications.

Regulatory Engagement and Performance Alignment

Navigating a rapidly evolving regulatory landscape is another key responsibility of Sustainability Consultants. Their role includes interpreting, applying, and communicating relevant environmental regulations and guidance documents, ensuring that performance targets not only meet compliance, but contribute to long-term sustainability goals.

Key Regulations and Guidance Areas:

- **Part L (Conservation of Fuel and Power):** Consultants ensure that dynamic thermal models demonstrate compliance with not just current regulations but also anticipated future standards, such as the Future Homes and Future Buildings Standards. This includes ensuring compliance with notional building benchmarks, Target Emission Rates (TER), and Fabric Energy Efficiency (FEE) targets.

- **Part F (Ventilation):** Consultants collaborate with M&E designers to ensure that natural and mechanical ventilation strategies deliver adequate airflow while minimising energy use and CO_2 levels. They also support Part F testing protocols post-construction.
- **Part O (Overheating Mitigation):** For residential buildings, consultants use TM59 or dynamic simulation to assess overheating risk and develop passive design solutions (e.g. external shading, orientation, thermal mass) that satisfy Part O compliance.
- **London Plan & GLA Energy Hierarchy:** In Greater London, consultants prepare Energy Statements demonstrating how the project meets the "Be Lean, Be Clean, Be Green" hierarchy. This involves modelling baseline and improved energy performance, and often, Whole Life Carbon (WLC) and Circular Economy Statements, as part of planning submissions.
- **Planning Conditions and S106 Agreements:** Consultants support planning compliance by ensuring that BREEAM, biodiversity net gain, and carbon offset requirements (e.g. financial contributions) are appropriately incorporated into the design and construction strategy.

Performance-Based Contribution: Rather than treating regulations as "tick boxes," Sustainability Consultants integrate them into a wider, outcome-focused framework. Their goal is not just compliance, but excellence, future-proofing the building against tightening legislation and aligning with client ESG, net-zero, and resilience strategies.

Conclusion: Orchestrating Performance Through Collaboration

Interdisciplinary collaboration is the engine of performance-based sustainability, and Sustainability Consultants serve as its conductor. By integrating data, facilitating communication, challenging assumptions, and aligning technical and strategic decisions with environmental objectives, they enable the team to deliver outcomes that are:

- Technically sound
- Cost-effective
- Compliant with regulations
- Optimised for long-term performance

As regulations tighten, client expectations evolve, and climate imperatives grow more urgent, the role of Sustainability Consultants in coordinating sustainable decision-making across disciplines will become ever more central to delivering buildings that are truly high-performing, resilient, and future-ready.

Post-Occupancy Evaluation and Operational Monitoring

A cornerstone of performance-based sustainability is validating that a building performs in reality as predicted during design. While energy models, environmental simulations, and carefully engineered systems set the theoretical performance parameters, true success is measured in post-construction operation.

This verification phase is critical to ensure that energy use, indoor environmental quality (IEQ), and occupant satisfaction align with both the project's environmental aspirations and its regulatory or certification commitments.

Sustainability Consultants lead this pivotal final phase through Post-Occupancy Evaluation (POE) and Operational Monitoring, closing the loop between predicted and actual performance. These processes also form the feedback engine of continuous improvement across building portfolios and inform future designs, retrofit strategies, and organisational sustainability policy.

Post-Occupancy Evaluation (POE)

Post-Occupancy Evaluation is a structured, evidence-based process that investigates how a building functions in real use, typically undertaken between 6 to 18 months after occupation. This allows systems and users to stabilise while still being early enough to identify and rectify issues before they become embedded.

Sustainability Consultants design and deliver POE studies that combine:

- Quantitative monitoring (e.g. metered energy, temperature, CO_2, IAQ)
- Qualitative feedback (e.g. occupant surveys, interviews, walkthroughs)

Key Evaluation Domains:

- **Energy Performance:** Comparison of actual consumption data (via sub-meters or utility bills) against simulated energy models (e.g. SBEM, EnergyPlus, IESVE) to identify performance gaps. This includes regulated (e.g., lighting, HVAC) and unregulated (e.g., plug loads) energy.
- **Thermal Comfort:** Assessed using a combination of sensor data and CIBSE TM52/TM59 criteria, along with occupant comfort feedback to understand heat distribution, draughts, and system effectiveness.
- **Indoor Air Quality (IAQ):** Monitored for CO_2 concentration, VOCs, PM2.5, and relative humidity. These results are evaluated against standards such as WELL, RESET, or BREEAM to determine if ventilation strategies and filtration systems are effective.

- **Lighting and Acoustics:** Measurement of lux levels, glare, circadian lighting control (where relevant), and acoustic separation in sensitive environments (e.g. education or healthcare) to support occupant wellbeing.
- **Occupant Satisfaction:** Captured using tools like the BUS Methodology, PROBE, or bespoke digital surveys to gauge user comfort, productivity, usability of systems, and perceived environmental quality.

Outcomes and Benefits:

- Identification of performance gaps and their root causes
- Insight into user experience and behavioural factors
- Validation of sustainability and energy claims
- Compliance with certification or ESG reporting obligations (e.g. WELL, BREEAM In-Use, GRESB)
- Informing future building strategies through lessons learned

The results are synthesised into a comprehensive POE report, often including charts, heat maps, and cross-references to simulation predictions. This report may also act as evidence for investors, planning authorities, or internal ESG teams.

Monitoring Systems and Building Analytics

Robust post-occupancy monitoring is a key enabler of operational excellence. It provides real-time insights into performance trends, enables early detection of faults, and supports long-term optimisation of systems.

Sustainability Consultants advise on and specify monitoring infrastructure that aligns with performance goals and certification standards.

Core Systems Include:

- **Advanced Sub-Metering:** Installed at the circuit or end-use level (lighting, HVAC, hot water, sockets), enabling disaggregated energy use analysis. Often integrated with BMS or cloud analytics platforms.
- **IEQ Sensors:** Monitoring CO_2, RH, temperature, VOCs, and particulate matter across different building zones. Some platforms also integrate occupancy sensors to evaluate space utilisation and adaptive comfort.
- **Building Management Systems (BMS):** Centralised control and analytics platform for HVAC, lighting, and other systems. Consultants advise on dashboard customisation, data granularity, and KPI tracking.

- **Energy Dashboards:** Public-facing interfaces or internal reports that visualise building performance, promoting transparency, behavioural engagement, and green credentials.
- **Digital Twins:** Virtual models that mirror the building's physical systems in real time. These enable advanced simulations (e.g., predictive maintenance, failure modelling) and support long-term retrofit planning.

Performance Integration:

- Performance thresholds are set for key indicators (e.g., <450 ppm CO_2, <12 $kWh/m^2/yr$ lighting energy).
- Alerts and diagnostics are programmed to flag deviations or inefficient operation.
- Data supports compliance with Net Zero verification, Streamlined Energy and Carbon Reporting (SECR), and internal sustainability reporting.

Continuous Commissioning and Operational Optimisation

Unlike one-off seasonal commissioning, continuous commissioning (Cx) refers to the ongoing adjustment and fine-tuning of building systems to maintain optimal performance.

Sustainability Consultants guide this process by:

- Scheduling regular system reviews to evaluate sensor accuracy, system response times, and sequencing logic
- Rebalancing air and water systems, particularly in buildings with variable occupancy, hybrid ventilation, or weather-dependent strategies
- Reprogramming control setpoints, such as deadbands, time schedules, and seasonal setpoint resets, based on actual performance data
- Training FM teams in energy literacy and BMS navigation, reducing dependency on specialist interventions

Advanced commissioning processes may include:

- Soft Landings Framework (BSRIA) to support handover and stakeholder engagement
- Measurement & Verification Plans (M&V) aligned with IPMVP protocols
- KPI dashboards to track performance against defined targets (e.g. NABERS, GRESB, internal ESG metrics)

Continuous commissioning is especially important in buildings targeting:

- Net Zero Operational Carbon
- BREEAM In-Use or NABERS UK ratings
- Energy Performance Contracts (EPCs)

Reporting, Feedback Loops, and Organisational Learning

Post-occupancy insights have limited value unless they're acted upon, shared, and embedded into future decision-making. Sustainability Consultants prepare bespoke reports for multiple audiences, translating technical data into actionable guidance.

Stakeholder Reporting Includes:

- **Building Owners/Clients:** Performance outcomes vs. design intent; lifecycle cost and carbon projections; strategic retrofit planning
- **Facilities Managers:** Control tuning, maintenance triggers, system reliability insights
- **Design Teams and Developers:** Feedback loops for CPD sessions, post-project reviews, and iterative design refinement
- **Certification Bodies:** Evidence for schemes like BREEAM In-Use, WELL Performance Verification, or RESET Air
- **ESG/Investment Teams:** Reporting aligned to GRESB, SECR, Science-Based Targets, or Green Lease Agreements

This process transforms POE and monitoring into a platform for continuous improvement, not just for the building in question, but across entire portfolios and future projects.

Conclusion: From Predictions to Proven Performance

Post-Occupancy Evaluation and Operational Monitoring are where the theory of sustainable design meets the test of reality. Performance-based design is not complete until outcomes are measured, lessons are learned, and systems are optimised.

Sustainability Consultants play a vital role in:

- Validating performance claims
- Enhancing occupant wellbeing
- Supporting adaptive, resilient building operations
- Driving institutional knowledge and future-ready design

This ongoing commitment to feedback, analytics, and optimisation ensures that sustainability moves beyond ambition, becoming measurable, evidence-based, and continually improving. It is the foundation upon which truly high-performing, future-proofed buildings are built.

Conclusion: Enabling Long-Term Sustainability Through Performance Metrics

Sustainability Consultants are the strategic linchpins of performance-based building design (PBBD). Far beyond simply offering advice on energy use or material choices, they embed quantitative rigour, environmental foresight, and technical innovation into every phase of the building lifecycle. Their role ensures that sustainability is not merely a box to tick or a marketing aspiration—but a core design driver with measurable outcomes, continuous feedback, and enduring operational value.

In an era defined by climate volatility, rising energy costs, health crises, and shifting social expectations, the contribution of Sustainability Consultants is no longer optional—it is essential. Their presence within the design and delivery team is what transforms sustainability from a philosophy into a strategy, and from a target into a traceable performance profile.

From Intent to Evidence: Anchoring Design in Metrics

Through the use of simulation tools, lifecycle assessments, modelling platforms, and certification frameworks, Sustainability Consultants convert abstract goals, like carbon neutrality, wellbeing, and climate resilience, into defined performance metrics. These metrics are not theoretical, they are:

- Simulated during design (e.g., energy, daylight, IAQ, thermal comfort)
- Verified through certification (e.g., BREEAM, LEED, WELL, Net-Zero)
- Measured in real-time (e.g., POE, sub-metering, environmental sensors)
- Refined post-occupancy (e.g., continuous commissioning, feedback loops)

This data-driven process ensures that sustainability is transparent, accountable, and scalable, giving clients and stakeholders the confidence that environmental claims are grounded in fact, not intention.

Aligning with Broader Goals: Climate, Health, Equity, and Economics

Sustainability Consultants operate at the intersection of environmental science, regulatory compliance, and project economics. Their work supports broader societal agendas, including:

- **Climate Action:** Enabling operational and embodied carbon reductions aligned with the UK Net Zero Carbon Buildings Standard, RIBA 2030 Climate Challenge, and international 1.5°C targets.
- **Health and Wellbeing:** Ensuring indoor environments support cognitive performance, mental health, and physiological wellness—particularly in homes, offices, schools, and hospitals, through the use of WELL, RESET, and POE data.
- **Social Value and Equity:** Guiding decisions around accessibility, affordability, thermal equity, and community engagement, particularly in public and mixed-use developments.
- **Economic Performance:** Aligning low-carbon and passive design strategies with lifecycle cost savings, return on investment, and ESG-based asset valuation.

In doing so, they position sustainable buildings not as niche exceptions, but as mainstream, future-proof investments.

Sustainability as a Lifecycle Journey, Not a Design Moment

Perhaps most critically, Sustainability Consultants advocate for whole-life sustainability. They ensure that performance is not confined to a model or certificate but evolves over time. This includes:

- Establishing realistic, monitored performance targets at briefing stage
- Supporting interdisciplinary collaboration throughout design
- Verifying outcomes through certification and post-occupancy evaluation
- Empowering facilities teams and users through data, training, and feedback

Their involvement ensures that the design narrative extends beyond handover, into commissioning, operation, and eventual retrofit or reuse, thereby maximising value and minimising regret.

Shaping the Future of the Built Environment

As the built environment sector moves toward mandatory climate disclosures, Net Zero performance targets, and health-led design policies, Sustainability Consultants are increasingly positioned as translators between ambition and delivery. They help:

- Navigate complex and evolving regulations
- Integrate circular economy principles
- Enable innovation in passive and renewable systems
- Drive digital transformation through smart technologies and digital twins

By embedding performance metrics into design decisions, they future-proof projects against obsolescence and anchor them to long-term value, resilience, and societal benefit.

In Summary

Sustainability Consultants empower design teams and clients to achieve meaningful, measurable, and lasting sustainability outcomes. Through simulations, certifications, lifecycle tools, collaborative leadership, and operational insight, they guide projects toward buildings that are:

- Environmentally restorative
- Economically viable
- Technically sound
- Socially and physically healthy
- Resilient across time, climate, and use

In performance-based building design, their role is not supplemental, it is foundational. It is their commitment to data, dialogue, and durability that ensures buildings are not just sustainable in name, but in performance, and that they continue to deliver value long after the ribbon is cut.

Chapter 8

Role of Cost Consultants / Quantity Surveyors in Performance-Based Building Design

In the evolving landscape of modern construction, performance-based building design (PBBD) presents a dynamic and goal-oriented framework that demands careful alignment of functional objectives with cost-effectiveness and long-term value. Within this paradigm, Cost Consultants and Quantity Surveyors (QS) play a critical and increasingly strategic role. Their expertise is no longer confined to traditional cost estimation or contractual management; instead, they are pivotal in shaping how financial decisions drive and support the delivery of high-performance, sustainable, and resilient buildings.

At its core, performance-based design shifts the focus from prescriptive compliance to the achievement of specific, quantifiable outcomes, be it energy efficiency, fire safety, structural resilience, or occupant well-being. This shift requires a new level of financial intelligence and forecasting accuracy that QS professionals are uniquely positioned to provide. Their contribution begins at the earliest stages of a project, often before any design is committed to paper, and continues throughout the lifecycle of the building, extending into post-occupancy performance assessment and asset management.

Cost Consultants and QS professionals bring rigorous cost management frameworks to the inherently complex and often iterative nature of performance-based projects. They conduct feasibility studies that assess not only capital expenditure (CapEx) but also operational and maintenance costs (OpEx), replacement cycles, and end-of-life considerations. These analyses help clients understand the total cost of ownership (TCO) and make informed decisions that consider both short-term affordability and long-term financial performance.

In PBBD, where innovative materials, systems, and design strategies may fall outside traditional benchmarks, QSs provide essential economic insight into the value proposition of emerging solutions. Whether evaluating the financial viability of smart building technologies, comparing passive design features to mechanical systems, or assessing the return on investment of zero-carbon strategies, they ensure that cost decisions are aligned with performance outcomes. This involves not only assessing cost efficiency but also integrating broader concepts such as value engineering, life cycle costing, and cost-benefit analysis to guide performance-informed financial decision-making.

Furthermore, the QS serves as a critical interface between the design team, project management, sustainability consultants, and clients, translating complex performance criteria into clear financial strategies. Their work supports key project functions including budget setting, performance benchmarking, procurement strategy, and financial risk assessment. By proactively managing the economic impact of design alternatives, QSs enable clients to make financially sound choices that align with both budget constraints and performance aspirations.

As buildings become more intelligent, responsive, and environmentally accountable, the role of the Cost Consultant or Quantity Surveyor becomes increasingly intertwined with performance evaluation and monitoring. They contribute to post-occupancy evaluations (POEs) by analysing how actual performance metrics align with original cost predictions and design intents. This feedback loop not only validates investment decisions but also informs future performance-based projects.

In summary, in the context of performance-based building design, Cost Consultants and Quantity Surveyors are indispensable partners in achieving financial and functional excellence. Their ability to merge cost discipline with performance ambition ensures that buildings are not only fit for purpose and regulation but also financially resilient and future ready. Their expanded role represents a fundamental shift, from cost controllers to value enablers, helping to redefine how buildings are conceived, delivered, and sustained over time.

Managing Capital and Lifecycle Costs

One of the most critical and evolving roles of the Quantity Surveyor (QS) in performance-based building design (PBBD) is the comprehensive management of both capital expenditure (CapEx) and lifecycle costs (LCC). As performance-based approaches move away from prescriptive design rules toward outcome-focused objectives, the financial planning process must also evolve. It must accommodate the complex relationship between design choices, performance targets, material lifespans, maintenance regimes, energy consumption, and user satisfaction, all of which impact both initial costs and long-term financial viability.

This responsibility goes far beyond the traditional scope of budgeting, estimating, and value engineering. It represents a strategic, data-driven, and forward-looking financial discipline that supports innovation, mitigates risk, and enables long-term economic sustainability. Within this framework, the Quantity Surveyor becomes not just a cost controller but a key integrator of financial and technical performance metrics, translating complex engineering solutions and architectural concepts into tangible financial insights.

Capital Expenditure (CapEx) and Performance Ambition

At the outset of any PBBD project, the QS plays a crucial role in shaping the financial foundation for the entire lifecycle of the building. Early-stage cost plans are developed in parallel with the definition of performance targets, such as thermal efficiency, fire safety, acoustic comfort, or carbon neutrality. The QS evaluates how the selection of specific systems, materials, or technologies contributes to these outcomes and assigns accurate capital cost estimates.

Examples of capital cost-sensitive decisions include:

- Choosing between conventional and high-performance façade systems, with implications for insulation values, solar gain, and structural integration.
- Selecting low-carbon materials such as cross-laminated timber, which may have higher upfront costs but reduce embodied carbon and construction time.
- Opting for intelligent building systems that provide dynamic energy management, fire detection, or occupancy-based lighting control.

The QS evaluates these elements not in isolation, but as part of an integrated performance package. This includes costing alternative design scenarios, conducting sensitivity analyses, and ensuring that the capital investment aligns with both the client's financial capacity and the performance expectations of the project.

Moreover, they support design team collaboration, advising architects, engineers, and sustainability consultants on how to achieve performance targets efficiently maximising value without compromising ambition.

Lifecycle Costing (LCC): Anticipating the Long View

While CapEx represents the financial foundation of the project, the real value of a performance-based approach is often realised across the building's operational life. Lifecycle costing allows the QS to extend the cost evaluation process beyond construction and into the building's use, maintenance, adaptation, and eventual decommissioning.

Lifecycle Costing (LCC) captures:

- **Operational costs:** energy, water, and maintenance.
- **Replacement cycles:** for systems such as HVAC, roofing membranes, lighting, or lifts.

- **Downtime implications:** particularly critical in healthcare, data centres, or mission-critical facilities.
- **Degradation modelling:** reflecting how materials or systems lose efficiency over time.
- **Residual value and salvage:** at end-of-life or refurbishment.

For example, when assessing ventilation options, a QS might evaluate a high-performance heat recovery system against a standard mechanical ventilation unit. While the heat recovery system costs more to install, it may:

- Reduce operational energy costs by 35%
- Require less frequent servicing
- Qualify for green energy incentives
- Extend system life through reduced strain

This comparative analysis enables clients to see beyond short-term affordability and understand the total cost of ownership (TCO), a critical metric for informed, performance-aligned decision-making.

In performance-led projects, LCC also enables compliance with sustainability frameworks (e.g., BREEAM, LEED, or NABERS) that require lifecycle and embodied carbon reporting. By linking cost to performance and environmental impact, the QS becomes central to the project's sustainability narrative.

Capital vs. Lifecycle: A Balancing Act

An enduring challenge in performance-based building design is the balance between initial capital investment and long-term financial return. Quantity Surveyors play a critical role in helping clients navigate this balance, particularly when upfront expenditure may exceed traditional norms.

Key QS interventions include:

- **Value optimisation workshops:** to explore material substitutions, passive design strategies, or system consolidation.
- **Investment appraisals:** quantifying return on investment (ROI), net present value (NPV), and internal rate of return (IRR) of performance-enhancing upgrades.
- **Phasing and futureproofing strategies:** for example, shell-and-core installations of infrastructure that support later upgrades without major retrofits.

In doing so, QSs ensure that performance is delivered where it matters most, from both a cost-benefit and operational standpoint.

Advising Across the Building Lifecycle

The strategic role of the QS does not end at practical completion. In performance-based design, where operational efficiency is continuously monitored and adjusted, the QS may support:

- Post-occupancy evaluations (POE) by comparing actual operating costs to original predictions.
- Facilities management integration, ensuring LCC data informs maintenance planning and system optimisation.
- Feedback loops into future projects, refining cost models based on observed performance.

This cradle-to-grave cost intelligence supports asset managers, investors, and building operators in maintaining performance standards, managing budgets, and planning long-term refurbishment or disposal strategies.

Conclusion

In performance-based building design, managing capital and lifecycle costs is not just about achieving cost efficiency, it's about enabling buildings to function, perform, and endure in line with client expectations and environmental responsibilities. The Quantity Surveyor's role is central to this mission. Through rigorous cost planning, strategic lifecycle analysis, and performance-integrated financial modelling, QSs ensure that innovative design choices are economically sustainable and that every pound spent contributes to both present value and future performance.

They are the stewards of financial realism in a design environment increasingly driven by complexity, ambition, and transformation, bridging the gap between visionary outcomes and tangible, cost-effective delivery

Capital Cost Control

Capital cost control is a cornerstone responsibility of the Quantity Surveyor (QS) in any construction project. However, in the context of performance-based building design (PBBD), the nature of capital cost control becomes significantly more dynamic, strategic, and iterative. Unlike traditional, code-compliant designs, where costs are typically anchored to well-established specifications and prescriptive regulations, PBBD allows and often necessitates bespoke, project-specific solutions. These can vary widely in scale, complexity, and material selection, each carrying its own set of cost implications.

In this environment, QS professionals act as financial strategists, providing detailed and responsive cost forecasting that evolves in tandem with the design. Their role is not only to contain costs but to ensure that financial resources are optimally aligned with the project's performance objectives, be it environmental sustainability, occupant safety, structural robustness, or long-term operational efficiency.

Early-Stage Forecasting with Performance Integration

From the earliest conceptual stages, QSs work closely with design teams to model the capital cost implications of key performance drivers. Rather than applying historical unit rates or benchmarks in isolation, they develop performance-integrated cost models that take into account:

- Material selection based on U-values, embodied carbon, or durability
- System efficiency ratings (e.g. HVAC seasonal performance factor)
- Spatial and geometric performance outcomes (e.g. solar access, airflow paths)
- Resilience features (e.g. flood mitigation, seismic isolation)
- Enhanced fire safety elements (e.g. pressurised lobbies, smoke venting)

QSs are instrumental in mapping design intent to financial reality. For example, if a building is aiming for Passive House certification, the QS must assess the incremental costs associated with triple glazing, advanced insulation, thermal bridge mitigation, airtight construction, and mechanical ventilation with heat recovery (MVHR). Each of these elements contributes to performance outcomes, but their collective capital cost needs to be clearly communicated, justified, and strategically managed.

Modelling Diverse High-Performance Systems

In a performance-based context, designers have the flexibility to explore non-traditional and innovative systems, many of which lack conventional cost baselines. This may include:

- **Enhanced thermal performance systems**
  → Vacuum-insulated panels (VIPs), phase-change materials, or double-skin façades
- **Structural resilience technologies**
  → Ductile detailing, tuned mass dampers, shear walls with embedded sensors
- **Fire protection measures**
  → Compartmentation schemes, advanced smoke extraction, intumescent coatings

QSs evaluate these not just on their unit costs, but on integration, logistics, supply chain availability, installation time, and maintenance requirements, offering a comprehensive view of capital impact.

They also model cost-performance ratios, helping clients decide between multiple pathways to the same performance goal. For instance, enhanced thermal comfort could be achieved via better insulation or more responsive HVAC systems. The QS helps quantify the comparative value of each option.

Dynamic Cost Planning in an Iterative Design Environment

One of the defining characteristics of PBBD is its iterative design methodology, driven by feedback from simulation tools, digital modelling (e.g., CFD, daylight analysis), and stakeholder inputs. Capital cost planning in this environment must therefore be:

- **Flexible:** Ready to incorporate rapid changes in specification or geometry
- **Responsive:** Capable of evaluating real-time impacts from energy or structural models
- **Transparent:** Communicated clearly to the client and design team, with assumptions documented

QSs provide progressive cost plan updates at each design gateway, concept, developed design, technical design, so that emerging ideas are continuously assessed against budgetary constraints. This prevents value erosion and cost surprises later in the procurement process.

They also use digital tools such as 5D BIM (cost-integrated modelling) to instantly visualise the impact of design decisions. For example, increasing the window-to-wall ratio can improve daylighting but also increase heat gain and façade costs. BIM allows QSs to show this financial impact in real-time, empowering more informed design choices.

Avoiding Over-Specification and Enhancing Synergy

In performance-based projects, there is a risk that, in pursuing ambitious outcomes, design teams may over-specify systems, selecting high-cost solutions where simpler alternatives might suffice. QSs serve a critical function in rationalising specifications to ensure that performance goals are met without unnecessary expenditure.

This may involve:

- **Identifying redundancy:** for instance, using both mechanical and natural ventilation where one could suffice
- **Highlighting duplication:** such as specifying overly robust structural systems with excessive safety margins

- **Recommending integration:** e.g., using external shading to reduce internal cooling load, enabling downsizing of HVAC

By seeking interdisciplinary synergies, QSs help reduce duplication and leverage design features that serve multiple functions, such as green roofs that provide insulation, stormwater management, and biodiversity value.

This approach allows the project to maintain or even improve its performance outcomes while bringing capital costs back in line with the client's expectations or financial constraints.

Conclusion: Strategic and Adaptive Capital Cost Control

In performance-based building design, capital cost control is a strategic and adaptive process. It is not about minimising expenditure at all costs but about maximising value for performance. The Quantity Surveyor's role is to bridge the gap between aspirational design and financial viability, ensuring that every design iteration is underpinned by rigorous financial analysis.

By embracing dynamic forecasting tools, scenario modelling, and cross-disciplinary collaboration, QSs help to de-risk innovative solutions and enable high-performance outcomes to be delivered on budget, on time, and with measurable value.

In doing so, they reinforce the essential principle that cost is not merely a constraint, but a design parameter, a critical lens through which every performance decision must be viewed.

Lifecycle Costing (LCC): A Strategic Tool for Long-Term Value

In the realm of performance-based building design (PBBD), short-term thinking is not sufficient. The imperative to create buildings that are not only efficient at the point of delivery but remain functionally, economically, and environmentally sustainable over decades demands a more holistic financial methodology. This is where Lifecycle Costing (LCC) becomes a powerful tool in the hands of the Quantity Surveyor (QS).

LCC enables the design team and client to look beyond initial capital investment and consider the total cost of ownership (TCO), including construction, operation, maintenance, refurbishment, and eventual disposal. It provides a structured, analytical framework to evaluate how design and specification decisions made today will financially impact a building's performance tomorrow, next decade, or even fifty years into its lifecycle.

Reframing Value: Beyond First Cost

In conventional procurement models, decisions are often driven by lowest initial cost, with limited consideration for future consequences. This can result in buildings that are cheaper to build but expensive to operate, maintain, or adapt.

In contrast, PBBD, particularly in projects targeting high performance standards such as net-zero energy, climate resilience, or low operational carbon, requires decisions that balance initial investment against enduring value. LCC empowers QSs to challenge first-cost thinking by showing how investments in better-performing systems or materials can yield:

- Lower operating costs (e.g., energy and water consumption)
- Fewer maintenance interventions (e.g., self-cleaning façades, robust floor finishes)
- Longer replacement intervals (e.g., LED lighting with 50,000+ hour lifespan)
- Greater resilience to changing regulations or energy prices
- Enhanced user satisfaction and asset value

These insights enable the QS to act as both financial analyst and value advocate, helping clients make rational decisions that align with their strategic, environmental, and operational goals.

Practical Applications of LCC in PBBD

A classic application of LCC is in mechanical systems selection. Consider two HVAC options:

- **Option A:** A basic split system with low upfront cost but poor energy efficiency, frequent maintenance needs, and a 10-year expected lifespan.
- **Option B:** A high-performance variable refrigerant flow (VRF) system with integrated heat recovery and intelligent controls. It costs 30% more to install but:
    - Reduces annual energy use by up to 40%
    - Requires servicing only every 2–3 years
    - Has a lifespan of 20+ years with fewer breakdowns
    - Lowers carbon emissions, aiding in achieving ESG targets or green certification

The QS models not just the installation cost, but:

- Annual operating costs over the building's life
- Discounted cash flows to account for time value of money

- Replacement and disposal costs
- Residual value, where applicable
- Non-financial benefits, such as increased indoor air quality or tenant comfort

Through metrics such as Net Present Value (NPV), Internal Rate of Return (IRR), and Simple Payback Period, the QS provides stakeholders with a robust, comparative economic picture. These analyses support a justified business case for higher-performance choices, particularly when operating budgets or carbon reduction targets are long-term priorities.

Incorporating Risk-Adjusted Costing

One of the most powerful aspects of LCC in PBBD is the ability to integrate risk-adjusted cost forecasting. This adds a layer of probabilistic thinking to the financial model, helping account for uncertainties such as:

- Equipment failure rates and associated downtime
- Unexpected repair costs or part unavailability
- Regulatory changes (e.g., carbon taxes, stricter energy codes)
- Energy price volatility
- Climate-related risks (e.g., flood or heat stress impacts on systems)

For example, a QS assessing the use of air-cooled chillers versus ground source heat pumps may factor in:

- The increased exposure to overheating risk in urban environments
- The escalating cost of grid electricity versus renewable sourcing
- The financial penalties of failing to meet performance-based maintenance or emissions targets

These risk elements are often included in Monte Carlo simulations, sensitivity analyses, or contingency modelling, giving clients a clearer understanding of best-case, worst-case, and most-likely financial trajectories over time.

LCC and Sustainability Reporting

As clients face mounting pressures to meet Environmental, Social, and Governance (ESG) goals, lifecycle costing becomes instrumental in quantifying the economic case for sustainability. LCC outputs can be directly linked to:

- Whole-life carbon assessments
- Green building certification criteria (e.g., BREEAM, LEED, WELL)
- Corporate sustainability reporting frameworks (e.g., GRESB, CDP, TCFD)

By integrating cost and carbon data, QSs help position the project within a triple-bottom-line framework, economic, environmental, and social value, providing a compelling narrative to funders, insurers, investors, and users.

Conclusion: Long-Term Thinking for Long-Term Value

Lifecycle Costing (LCC) transforms the role of the Quantity Surveyor from a short-term cost estimator to a long-term financial strategist, empowering decision-makers to invest wisely in performance-driven solutions. In performance-based building design, where innovation, flexibility, and resilience are paramount, LCC ensures that the full financial trajectory of every system and material is visible, analysable, and optimised.

It provides the tools to answer critical questions: *Is this investment financially sustainable? Will it reduce long-term risk? Is it aligned with our environmental obligations?* In doing so, LCC becomes not just a method, but a mindset, one that drives high-value, high-performance, and future-proofed building outcomes.

Whole Life Value: Maximising Long-Term Performance and Investment Return

In the context of performance-based building design (PBBD), decision-making must extend far beyond short-term affordability and isolated financial metrics. Buildings today are increasingly viewed not only as physical assets but as long-term investments with complex value profiles, including environmental impact, operational resilience, adaptability, and user wellbeing. To support this integrated outlook, Quantity Surveyors (QSs) employ a more expansive and strategic financial methodology known as Whole Life Value.

Whole life value assessment takes a holistic view of a building's financial and performance journey from inception through to end-of-life, encompassing all aspects of cost, benefit, risk, and residual potential. It combines upfront capital expenditure, lifecycle operating costs, environmental stewardship, and social responsibility into a single, unified framework. By doing so, it allows decision-makers to make informed, long-term choices that align with economic, environmental, and institutional priorities.

Key Dimensions of Whole Life Value Analysis

	Metric	Description
1	Upfront Capital Costs	These include the design, procurement, and construction expenses associated with delivering the building to performance standards. While high-performance systems and resilient materials may increase initial costs, whole life value analysis assesses whether those investments provide net benefits over time.
2	Operational and Maintenance Costs	This considers the building's performance during occupancy—energy consumption, water usage, cleaning, system servicing, and routine maintenance. A building with smart systems and optimised controls may reduce energy bills and maintenance frequency, improving operational cost efficiency.
3	Scheduled Replacements and Refurbishments	QSs develop elemental replacement schedules to forecast when and how often building systems, such as HVAC, cladding, or electrical distribution, will require upgrade or overhaul. This allows clients to anticipate capital reinvestment needs over decades, not years.
4	Residual Value and End-of-Life Disposal	At the end of its functional life, a building (or its components) may retain salvageable or reusable value. Modular systems, recyclable materials, and demountable structures can improve financial and environmental outcomes through repurposing or resale.
5	Environmental and Social Value Contributions	Whole life value goes beyond economic metrics to include the building's carbon footprint, climate resilience, occupant health, accessibility, and community benefits. These are increasingly important in justifying public funding, private investment, and regulatory approvals.

Applications in Sustainability and Resilience Strategies

Whole life value assessment is especially relevant in mission-critical and public-facing buildings, such as hospitals, schools, and civic infrastructure, where long-term performance, reliability, and social utility are paramount.

For example:

- In designing a hospital, QSs may recommend reinforced structural systems, flood barriers, or uninterrupted power supply (UPS) installations to ensure business continuity in extreme events. Though costly up front, the long-term savings from avoided downtime and reputational risk are substantial.
- In a school project, the use of thermally massive construction and natural ventilation can lower cooling demands and reduce future refurbishment needs. The associated social value, improved learning environments, lower running costs for public bodies, further elevates the return on investment.

By modelling the interplay between physical durability, operational efficiency, and performance resilience, the QS ensures that the building is not only functional and cost-efficient today, but capable of adapting to future needs and stresses.

Advanced Cost-in-Use Modelling and Material Analysis

QSs use a variety of tools to deliver whole life value assessments, including:

- **Cost-in-use modelling:** Predicting annual costs over decades and factoring in inflation, degradation, and escalation scenarios.
- **Replacement forecasting:** Identifying lifecycle stages of materials and systems and aligning those with anticipated building use and renewal cycles.
- **Durability and recyclability analysis:** Comparing materials not just for cost and performance, but for their ability to withstand wear, be repurposed, or avoid landfill.

For example, in the selection of external cladding, the QS may analyse:

- Aluminium rainscreen panels (durable, recyclable, high capital cost)
- Fibre cement boards (lower cost, moderate durability, minimal resale value)
- Timber siding (biogenic carbon storage, shorter lifespan, potentially higher maintenance)

These insights are not presented in isolation but within the context of project-specific goals e.g., targeting BREEAM Excellent, LEED Platinum, or alignment with a corporate net-zero pathway.

Integrating LCC and Whole Life Carbon Metrics

In progressive projects, QSs combine Lifecycle Costing (LCC) with Whole-Life Carbon (WLC) assessment, creating a powerful toolset for:

- Quantifying cost per tonne of carbon saved
- Modelling carbon pricing risk exposure over time
- Demonstrating the return on investment of low-carbon and renewable technologies (e.g., photovoltaic panels, CLT, green roofs)

For instance:

- A high-performance window system may cost 25% more initially but deliver a 60% reduction in heat loss and reduce carbon emissions significantly.
- A green roof may provide stormwater retention, insulation, and biodiversity enhancements, each with social and environmental value multipliers.

This combined analysis provides a, financial, environmental, and operational summary, enabling clients to align building investments with ESG reporting frameworks and long-term strategic goals.

Conclusion: Whole Life Value as a Foundation for Informed Decision-Making

In performance-based building design, Whole Life Value is not an optional add-on, it is a foundational principle. It acknowledges that buildings are not momentary products, but enduring systems with economic, environmental, and societal consequences.

Through detailed forecasting, material analysis, and integrated performance-cost modelling, Quantity Surveyors provide clarity amid complexity. They help clients see not just the cost of building today, but the value of owning, operating, maintaining, and evolving it tomorrow.

By leveraging Whole Life Value thinking, projects are better equipped to deliver resilient, efficient, low-carbon, and user-centred outcomes, ensuring that every investment decision adds meaningful, measurable, and lasting value over the building's entire lifespan.

Conclusion

In performance-based building design, effective management of capital and lifecycle costs is essential to delivering value-driven, future-ready buildings. Through a combination of traditional cost control, advanced lifecycle analysis, and whole life value assessments, Quantity Surveyors empower clients to make decisions that optimise both financial outcomes and building performance. Their expertise ensures that innovative solutions are economically viable, environmentally sound, and strategically aligned with the project's long-term operational vision.

Economic Impact of Design Decisions

In performance-based building design (PBBD), each design choice is tied not only to functional or aesthetic outcomes but to quantifiable performance targets, such as energy use, carbon emissions, fire resilience, or structural robustness. These decisions often require the integration of advanced technologies, innovative materials, or alternative construction methods that may significantly deviate from traditional benchmarks. Within this complex environment, Cost Consultants and Quantity Surveyors (QSs) provide essential financial clarity, helping stakeholders navigate the economic consequences of performance-driven choices.

Their role is to enable financial transparency and informed decision-making, ensuring that the economic viability of various options is rigorously tested against their projected performance. This involves not only assessing direct costs, but also considering longer-term financial outcomes, risks, incentives, and opportunity costs.

Evaluating Performance Trade-Offs

One of the most significant contributions QSs make to PBBD is their ability to conduct cost-performance trade-off evaluations. Because performance-based design is inherently flexible, allowing design teams to choose from a range of strategies to meet a given performance goal, there are often multiple valid solutions, each with distinct cost profiles and implementation risks.

For example:

- **Thermal Performance:** Should the design include triple glazing with low-emissivity coatings or rely on dynamic external shading systems? Both strategies may achieve similar energy performance, but their capital costs, maintenance implications, and operational reliability differ.
- **Structural Fire Resistance:** Is cross-laminated timber (CLT) a cost-effective and code-compliant alternative to concrete in multi-storey buildings? While CLT may reduce construction timelines and embodied carbon, the QS must account for fire protection costs, insurance implications, and additional regulatory compliance efforts.
- **Acoustic vs Ventilation Trade-offs:** In urban high-rises, achieving good natural ventilation may conflict with acoustic performance. QSs assist in

determining whether investment in hybrid ventilation with noise-attenuating features represents better value than mechanical-only systems.

These trade-off studies help clients and design teams optimise resource allocation, ensuring that performance targets are met in the most economically efficient way.

Scenario Analysis

To support strategic decision-making, QSs frequently perform scenario-based economic assessments. These scenarios may reflect:

- Different performance levels (e.g., meeting minimum code vs achieving LEED Platinum or BREEAM Outstanding)
- Varying operational assumptions (e.g., occupancy density, energy prices, climate conditions)
- Alternative construction methods or phasing strategies

For instance:

- A public building project may consider two options:
    - a base-case design meeting minimum regulatory requirements for energy performance and
    - a performance-optimised design targeting net-zero operational carbon. The QS compares the incremental capital cost of high-performance envelope systems, solar PV arrays, and smart controls against the lifecycle energy cost savings and potential income from feed-in tariffs or renewable energy credits.
- In a flood-prone site, the design team may propose either flood-resilient construction (e.g. raised foundations, sacrificial ground floors) or the installation of permanent perimeter defences. QSs calculate the capital cost differential and assess the economic risk of flooding, including insurance premiums, business disruption, and repair costs.
- Scenario analysis also allows QSs to test the sensitivity of decisions to changes in external variables, such as energy tariffs, carbon pricing, or regulatory tightening, providing clients with a robust understanding of financial exposure under future conditions.

Incentive and Grant Advising

As sustainability and resilience rise on both political and corporate agendas, governments and institutions are offering a growing range of financial incentives, subsidies, and regulatory credits to encourage performance-based improvements.

These incentives can significantly alter the cost-benefit equation of high-performance solutions.

Cost Consultants play an essential role in:

- **Identifying applicable incentives:** QSs stay up to date with local, national, and international programs that may support performance investments, such as:
    - Green building grants
    - Renewable energy subsidies (e.g. for heat pumps, solar panels)
    - Enhanced capital allowances (e.g. for energy-efficient systems)
    - Reduced VAT or tax credits for sustainable construction
- **Quantifying financial benefit:** They evaluate how incentives reduce net capital expenditure or improve payback periods. For instance, a high-efficiency HVAC system eligible for a government rebate might shift from being marginally cost-effective to clearly advantageous.
- **Ensuring compliance and claim ability:** Many programs require formal application, documentation, and ongoing verification. QSs coordinate with sustainability consultants to ensure that the design and procurement processes meet the criteria necessary to secure funding.

In some cases, incentive packages can be aggregated or layered, meaning that QSs must develop a comprehensive financial strategy to maximise value extraction. This is especially critical in large-scale public or mixed-use developments, where multiple funding streams, from environmental grants to social infrastructure incentives, may be available.

Conclusion

The economic impact of design decisions in PBBD is complex, nuanced, and multi-dimensional. Quantity Surveyors act as economic navigators, ensuring that design innovation is matched by financial viability. Through performance trade-off evaluations, scenario modelling, and grant advisory, they empower clients to choose design pathways that not only meet technical and environmental goals but also align with broader financial, regulatory, and investment strategies. Their ability to link performance outcomes with financial foresight is a cornerstone of successful performance-based building projects.

Supporting Cost-Benefit Analysis of Performance-Based Alternatives

This section examines how Quantity Surveyors (QS) and Cost Consultants evaluate the economic implications of design decisions in performance-based building design

(PBBD). It focuses on cost-performance trade-offs, scenario analysis, and leveraging incentives and grants. The included chart and case studies demonstrate how financial strategies are applied to support sustainable, resilient, and high-performing buildings.

Case Studies

CLT vs Concrete in Multi-Storey Residential Project

In a mid-rise residential project, the design team evaluated the use of cross-laminated timber (CLT) instead of conventional concrete. The QS conducted a cost-performance trade-off analysis comparing material costs, fire resistance treatments, insurance premiums, and construction speed. Although CLT incurred a 12% higher capital cost, faster assembly led to significant program savings and lower site labour expenses. Further, embodied carbon reductions aligned with the client's ESG goals, unlocking green finance incentives and improving the long-term investment profile.

Net-Zero Public Library – Scenario Analysis for Envelope Design

In the early stages of a new public library, the project team developed multiple envelope performance scenarios. The QS analysed three levels of façade insulation and glazing quality, ranging from building regulation minimums to Passive House standards. Cost assessments considered not only material premiums but also long-term energy savings and available government grants. The mid-range solution, featuring high-performance glazing and cavity insulation, offered the best value, with a 6-year payback and compliance with local carbon targets.

Incentive-Driven Energy System Upgrade in Mixed-Use Development

In a major urban regeneration project, the QS team explored the financial viability of integrating a centralised heat pump system backed by solar PV. Initial cost concerns were mitigated when the QS identified regional subsidies covering 25% of the renewable installation cost. Combined with long-term energy savings and operational carbon reductions, the incentive transformed the viability of the proposed system. The strategy not only supported the building's BREEAM Excellent rating but also enhanced marketability to ESG-focused tenants.

Enabling Transparent and Performance-Driven Budgeting

In the context of performance-based building design (PBBD), traditional budgeting methods, often based on fixed quantities and static pricing, are no longer sufficient. Instead, there is a growing need for dynamic, transparent, and performance-driven financial planning that evolves in parallel with the design process. As design decisions are increasingly shaped by performance simulations and predictive modelling, Quantity Surveyors (QSs) play a central role in ensuring that the associated financial implications are clearly understood, continuously tracked, and managed with precision.

QSs enable transparency by providing real-time cost visibility across all stages of a project and ensure that budgets remain aligned with evolving performance targets. This involves integrating cost planning with digital tools, applying data-driven benchmarking strategies, and maintaining vigilant oversight of both cost and performance metrics through to project completion.

Cost Planning Integrated with BIM and Digital Tools

The integration of Building Information Modelling (BIM) with cost planning has revolutionised how QSs engage with performance-based design. BIM enables the creation of a data-rich 3D model of the building that includes not only geometry but also specifications, material properties, and performance data. When linked with cost management software and databases, QSs can extract quantitative and qualitative information to generate highly accurate, real-time cost estimates.

In a PBBD framework, BIM integration allows QSs to:

- Dynamically update cost plans as design iterations occur, whether related to insulation thickness, glazing area, structural span, or renewable system sizing.
- Simulate cost impacts of performance changes, such as upgrading HVAC systems, increasing fire resistance, or improving structural damping.
- Align cost projections with simulation outcomes from energy modelling (e.g., EnergyPlus), lighting analysis (e.g., Dialux), or structural analysis tools.

For example, when an architect proposes a change to the façade system for improved thermal performance, the QS can immediately assess the cost delta, update the budget, and flag whether the change exceeds the allowable financial tolerance. This real-time responsiveness significantly enhances decision-making and reduces the risk of budget overruns later in the project.

Furthermore, BIM facilitates 4D (time) and 5D (cost) modelling, enabling QSs to visualise and forecast the financial implications of construction sequencing, procurement timing, and performance upgrades over time. This visibility supports proactive value engineering and scenario testing, rather than reactive cost-cutting.

Benchmarking Against Performance-Oriented Projects

Benchmarking plays a vital role in validating the financial feasibility of performance-driven strategies. By comparing costs, specifications, and outcomes against similar completed projects, QSs help stakeholders understand where their project stands in relation to current market norms and best practices.

In performance-based design, benchmarking is used to:

- Compare the cost efficiency of performance enhancements, such as comparing envelope upgrades across multiple zero-energy buildings.
- Gauge the market acceptability of capital investments, e.g. the use of smart lighting or climate-adaptive façades.
- Assess payback periods and return on investment against industry-wide trends.

QSs draw on databases such as BCIS (Building Cost Information Service), historical project records, government datasets (e.g., RIBA benchmarking), and proprietary performance-cost libraries. This data is contextualised for building type, location, climate zone, regulatory regime, and sustainability certifications.

Benchmarking also supports the development of performance-led cost targets. For example, in a commercial office building seeking BREEAM Outstanding certification, benchmarking may reveal that achieving a specific operational energy intensity (e.g., <100 kWh/m²/year) generally correlates with an envelope spend of 18–22% of total capital cost. These insights allow the QS to proactively allocate budget to high-impact areas, ensuring value for money in performance-driven decisions.

Monitoring Performance Goals Against Budgets

Once construction is underway, performance-based buildings require ongoing oversight to ensure that systems and materials installed reflect both the original performance criteria and the approved budget. QSs serve as financial gatekeepers during construction and commissioning, verifying that each component or system remains within budgeted thresholds while still delivering intended outcomes.

Key aspects of this oversight include:

- Reviewing submittals and procurement data to ensure performance-rated systems (e.g., fire dampers, heat recovery units, acoustic insulation) match specification and cost expectations.
- Tracking deviations from design intent during value engineering or contractor-led changes and advising whether proposed substitutions impact performance.

- Assessing commissioning outcomes, such as airflow, lighting levels, or thermal comfort data, and validating that they meet both contractual and performance requirements without incurring unplanned costs.

If discrepancies arise (e.g., a fire suppression system performs below expectations or a sustainable material is unavailable), QSs recommend cost-effective corrective actions, such as substituting with an equivalent system or reallocating contingency budgets. This ensures that performance compliance is not sacrificed due to unforeseen constraints or supplier issues.

Post-completion, QSs may also assist in reconciling performance outcomes with actual spend, generating reports that feed into post-occupancy evaluation (POE) processes and future benchmarking databases.

Conclusion

In performance-based building design, budgeting is not simply an exercise in cost control, it is an active, responsive, and integrated function that must support the project's performance vision. Quantity Surveyors enable this by embedding cost planning within digital tools like BIM, validating design decisions through industry benchmarking, and closely monitoring implementation to ensure budgets and performance targets remain aligned.

Their work ensures that the economic dimension of design remains transparent and justifiable, empowering stakeholders to make confident, data-driven decisions that uphold both financial integrity and performance excellence throughout the project lifecycle.

Enabling Vision Through Financial Precision

In the high-stakes, performance-driven landscape of modern construction, Cost Consultants and Quantity Surveyors (QSs) are more than just custodians of budgets—they are strategic enablers of design ambition. As the construction industry evolves to embrace higher standards of sustainability, resilience, adaptability, and occupant well-being, the QS's role becomes pivotal in bridging the gap between vision and feasibility.

In performance-based building design (PBBD), where outcomes, not prescriptions, define success, QSs provide the financial intelligence necessary to make informed, responsible decisions throughout the building lifecycle. They transform abstract performance goals into quantifiable financial strategies, ensuring that cost considerations are not viewed as constraints, but as tools to maximise value and performance.

Balancing Performance with Pragmatism

Whether evaluating options for thermal performance, structural resilience, carbon reduction, or user experience, QSs deliver cost-benefit clarity that helps project teams navigate a complex matrix of trade-offs. They are uniquely positioned to assess the true value of performance enhancements, not only in terms of capital investment but also in the context of:

- Long-term operational savings
- Risk reduction and resilience
- Lifecycle cost-efficiency
- Environmental impact
- Social and economic returns on investment

Their ability to analyse both quantitative data (NPV, IRR, CapEx, OpEx) and qualitative performance metrics empowers clients to select solutions that are not only affordable but also accountable to long-term expectations.

Enabling Informed, Future-Proofed Design

QSs are instrumental in creating buildings that are resilient, adaptable, and fit for the future. Through tools such as:

- Lifecycle Costing (LCC)
- Whole Life Value (WLV) analysis
- Performance-integrated cost planning
- BIM-linked 5D cost modelling
- Cost-risk forecasting

They ensure that design decisions consider not only the immediate needs of a project but its enduring relevance and sustainability. This long-range thinking is vital in a climate where economic volatility, environmental regulation, and societal expectations are all in flux.

Their input helps protect clients from short-sighted decision-making, underinvestment in critical systems, or over-specification that doesn't translate into measurable performance benefits.

Catalysts for Accountability and Transparency

QSs also play a vital role in transparency and stakeholder communication. In performance-based environments, where success is defined by whether specified outcomes are achieved, be it energy use intensity, fire resilience, acoustic comfort, or embodied carbon, cost consultants help:

- Track progress against budget and performance metrics
- Manage risk and contingency with foresight
- Document cost-performance decisions for accountability
- Align funding streams and incentives with design aspirations

Their ability to speak both the language of finance and performance makes them critical collaborators across interdisciplinary teams, including architects, engineers, sustainability consultants, facilities managers, and investors.

Sustainability, Value, and Innovation—Delivered Through Cost Discipline

In the drive toward zero-carbon buildings, smart infrastructure, and adaptive reuse, cost consultants ensure that innovation does not come at the expense of viability. They advocate for solutions that achieve multiple performance objectives with minimal financial redundancy, bringing a "do more with less" philosophy into the heart of design.

They also enable sustainability to be valued properly, quantifying the return on investment for green technologies, renewable systems, biophilic design, and other innovations that might otherwise be seen as non-essential.

Through cost-risk balance, lifecycle insight, and market benchmarking, they help realise buildings that are not only technically advanced but also financially robust, socially responsive, and environmentally responsible.

Final Reflection: From Cost Control to Value Creation

Ultimately, the contribution of Quantity Surveyors and Cost Consultants in PBBD represents a shift in mindset, from cost control to value creation. Their role is not to limit possibility, but to structure and empower it, ensuring that ambitious performance goals can be delivered efficiently, responsibly, and sustainably.

They are the stewards of long-term thinking in an industry that increasingly demands accountability, adaptability, and foresight. Their insight ensures that performance-based buildings are not only designed for excellence but also delivered for impact, economically, operationally, and ethically.

Chapter 9

Role of Building Surveyors in Performance-Based Building Design

As the construction industry continues to evolve toward more integrated, data-driven, and outcome-focused methodologies, Performance-Based Building Design (PBBD) has emerged as a transformative approach to defining and delivering building quality. In this context, success is no longer determined solely by adherence to prescriptive codes or minimum regulatory standards, but by the achievement of measurable performance objectives, such as thermal comfort, energy efficiency, structural resilience, fire safety, indoor environmental quality, and long-term durability. These objectives must be met not only at completion, but sustained throughout the entire lifecycle of the building.

In this paradigm, the role of the Building Surveyor is undergoing a profound and meaningful evolution. Traditionally seen as technical generalists responsible for identifying defects, managing remedial works, and overseeing compliance, Building Surveyors now occupy a far more strategic and interdisciplinary position in performance-led projects. They are emerging as critical custodians of performance assurance, ensuring that design intentions are accurately translated into built reality and that buildings continue to meet their intended functions in an efficient, safe, and sustainable manner.

Their function is unique in that it spans the full trajectory of the building process, from early-stage design support to construction oversight and post-occupancy evaluation. Operating at the confluence of design delivery, site execution, and lifecycle management, Building Surveyors provide the continuity, accountability, and technical rigour required to verify that a performance-based approach does not falter between concept and execution.

Crucially, Building Surveyors are also evolving into key interpreters of regulatory frameworks in non-prescriptive contexts. PBBD often involves innovative solutions, alternative materials, or non-standard methods that depart from established norms. In such cases, Building Surveyors help ensure that these innovations are robustly justified, technically substantiated, and legally defensible, engaging with statutory authorities, collating performance data, and ensuring that compliance is demonstrated through performance equivalency rather than prescriptive adherence.

Moreover, as buildings increasingly adopt complex systems and digital infrastructure, the role of the Building Surveyor includes responsibility for verifying installation quality, commissioning outcomes, and integration between performance-critical systems (e.g., HVAC, fire protection, access control, façade automation). Their oversight of quality assurance mechanisms, combined with a thorough understanding of both building fabric and system interfaces, positions them as essential contributors to the reliability and maintainability of performance-driven assets.

Perhaps most significantly, Building Surveyors in PBBD are now key players in lifecycle planning and asset performance optimisation. Their insight into material degradation, maintainability, access, and resilience enables them to advise on how design choices will impact operational cost, user safety, and long-term adaptability. By integrating their knowledge into asset registers, facilities handovers, and post-occupancy evaluations, they help ensure that buildings remain compliant, functional, and cost-effective over time, not just at handover, but across their entire lifecycle.

In summary, the role of the Building Surveyor in performance-based building design is no longer ancillary or reactive. It is central, proactive, and indispensable. These professionals serve as technical stewards, compliance strategists, quality enforcers, and lifecycle custodians, ensuring that performance aspirations are not only designed into buildings, but also built, validated, and sustained over decades. In an era where performance is paramount, Building Surveyors provide the assurance that vision meets reality, with measurable, lasting results.

Oversight of Site Inspections and Performance Verification

In performance-based building design (PBBD), site inspections take on a new level of technical significance and accountability. They are no longer routine walkthroughs or checklist-driven audits confined to surface-level compliance. Instead, they are structured, evidence-led processes of verification, integral to ensuring that what is constructed faithfully reflects the design intent and meets the defined performance objectives of the project.

At the heart of this process is the Building Surveyor, whose expertise enables the project team to bridge the gap between simulation and realisation, between specified performance outcomes and their physical, measurable achievement on site.

From Procedural to Performance-Oriented Inspections

Traditional building inspections are often centred around verifying compliance with prescriptive codes, confirming that materials and systems have been installed according to standard practice or that workmanship meets industry norms. In PBBD, this model evolves significantly.

Performance-based inspections shift the focus from prescriptive conformity to functional adequacy and outcome alignment. This requires a deep understanding of the intended performance criteria and a methodical approach to verifying how construction details, material application, and system integration contribute to those targets.

Examples include:

- Ensuring that thermal envelopes are installed without compromise, including proper insulation continuity, sealed vapour barriers, and elimination of thermal bridges—all validated against the energy model used in the design phase.
- Verifying that fire-resisting elements, such as shaft walls, service penetrations, and compartment lines, conform not just to drawings but to fire strategy documents and fire resistance performance ratings.
- Reviewing acoustic insulation installations against simulation results to ensure sound attenuation meets comfort or regulatory thresholds.
- Validating the physical configuration of ventilation openings, ductwork, or operable elements against CFD (computational fluid dynamics) models, especially in mixed-mode or passive systems.

Each of these inspections must be performed with an understanding that deviation from design, however minor, can result in significant performance degradation. Building Surveyors must therefore apply a critical eye and rely on documentary evidence, physical testing, and photographic records to substantiate their observations.

Evidence-Based Verification: The Core of PBBD Assurance

Unlike traditional projects that rely heavily on conformity to a fixed specification, PBBD allows for design flexibility, so long as performance outcomes are demonstrably achieved. This puts the onus on verifiability rather than prescriptiveness and it is here that the Building Surveyor's role becomes especially vital.

The verification process involves:

- Cross-referencing site conditions with simulation outputs, such as energy models, fire strategies, or daylighting studies
- Reviewing as-built conditions against manufacturer technical data, test certificates, and warranties to confirm stated performance claims
- Ensuring that materials and products used on site match those approved in submittals and performance declarations

- Identifying where site adaptations may necessitate performance re-validation or redesign, and liaising with designers to ensure compliance continuity

This process often demands collaboration with design consultants, commissioning agents, and systems engineers, ensuring that complex installations (e.g., high-performance glazing, active shading systems, low-energy HVAC) are not only installed correctly but integrated to perform collectively.

Examples of Performance Verification in Practice

To illustrate the nuance and rigor of PBBD inspections:

- In a Passivhaus-certified project, Building Surveyors verify the integrity of airtightness detailing—checking tapes, membranes, and junctions with calibrated testing equipment (e.g., blower door test). Their inspections support formal certification by identifying and remedying air leaks before testing.
- In a performance-led fire strategy, such as for an atrium building with smoke ventilation, the Building Surveyor ensures that system commissioning achieves the designed airflow rates and smoke clearance times, as defined in fire engineering calculations. Testing protocols and pass criteria are documented and signed off prior to occupancy.
- On a sustainable campus development, Building Surveyors may oversee the installation of natural ventilation systems, confirming actuator placement, daylight sensor alignment, and window operability to ensure that passive systems perform as modelled under real conditions.

Inspection Planning and Documentation

To ensure consistency and rigour, Building Surveyors in PBBD develop and implement:

- Performance-Focused Inspection Checklists tailored to individual project performance objectives
- Inspection Test Plans (ITPs) that define the method, frequency, and criteria for inspections across different trades and systems
- Non-Conformance Reporting (NCR) protocols with direct traceability to specific performance shortfalls
- Photographic and digital documentation to support audit trails, BIM integration, or commissioning records

Where performance is mission-critical (e.g., hospitals, labs, or clean rooms), these inspection regimes are further enhanced with commissioning witness tests, factory acceptance tests (FATs), and on-site validation procedures that allow Building Surveyors to sign off with confidence.

Conclusion: Ensuring Performance Is Built, Not Just Designed

In the world of performance-based building design, Building Surveyors are the linchpin between theoretical achievement and actual delivery. Their site inspections and verification processes ensure that every product, detail, and system is not only installed but installed correctly, consistently, and coherently with the building's intended performance profile.

By applying technical diligence, forensic attention to detail, and a collaborative ethos, Building Surveyors protect the integrity of performance-based outcomes, helping to deliver buildings that are not only code-compliant, but truly fit for purpose, future-ready, and capable of meeting the demands of time, use, and environment.

Quality Assurance and Construction Control: Safeguarding Performance Through Precision

In a performance-based building design (PBBD) framework, where buildings are evaluated not just on whether they meet a set of prescriptive criteria but on how well they achieve targeted performance outcomes, Quality Assurance (QA) becomes a critical mechanism of success. It is no longer limited to verifying workmanship or materials in isolation—it involves systematic oversight of the integration, calibration, and validation of performance-critical systems throughout the construction lifecycle.

Building Surveyors are at the heart of this expanded QA paradigm. Their expertise in construction detailing, regulatory compliance, system integration, and building science equips them to develop, implement, and monitor rigorous QA frameworks that uphold the integrity of performance goals—from air tightness and energy consumption to fire safety and structural robustness.

Beyond Workmanship: QA for Integrated Performance Systems

Modern performance-based buildings are highly complex, with numerous interdependent systems that must function in harmony to deliver intended results. These include, but are not limited to:

System	Description
Intelligent Building Management Systems (BMS)	Centralized digital platforms that monitor and control HVAC, lighting, security, shading, and energy metering. QA involves not only installation accuracy but ensuring that sensors, actuators, and interfaces are correctly programmed and tested to respond as intended to occupancy, environmental conditions, or user input.
High-Efficiency HVAC Installations	Systems designed for low-energy consumption and high thermal comfort must be sized, installed, and commissioned precisely. Building Surveyors check duct layouts, verify insulation, ensure zoning logic matches the energy model, and confirm airflow rates through commissioning records and performance verification.
Advanced Façade Systems	Whether it's operable louvres, smart glazing, or double-skin façades, these elements impact not only energy efficiency but occupant comfort and glare control. QA encompasses the installation tolerances, motorised controls, daylight response calibration, and water/wind resistance integrity testing.
Performance-Driven Structural Systems	Systems designed to support fire protection, seismic resilience, or long-span flexibility must be installed according to precise detailing. QA includes inspection of connection detailing, fireproofing application, anchoring systems, and checking conformity with structural engineer sign-offs.

Each of these systems must not only be installed but proven to function according to the performance criteria defined in the design documents and validated through simulation or analysis.

The Role of Building Surveyors in Quality Assurance Planning

Building Surveyors develop and administer Quality Assurance Plans (QAPs) tailored to the project's performance goals. These plans define:

- Inspection points and hold-points for critical systems and assemblies
- Performance testing requirements (e.g. thermal imaging, blower door tests, commissioning tests)

- Documentation protocols, including checklists, witness signatures, submittals, and photographic evidence
- Traceability frameworks that connect built outcomes back to simulation models, manufacturer specifications, and approval records

QAPs are structured documents that provide clear benchmarks and responsibilities for contractors, subcontractors, and third-party verifiers. They are living documents, updated throughout construction to reflect changes in design scope, regulatory feedback, or system substitutions.

In many projects, the Building Surveyor collaborates with the Commissioning Manager, Clerk of Works, and Design Team Lead to maintain a common performance verification process across disciplines.

Alignment with Performance Certification and Project Objectives

QA is especially critical where projects pursue formal performance certifications, such as:

- BREEAM, LEED, WELL, or Passivhaus
- Net Zero Energy (NZE) or Zero Carbon targets
- Resilience frameworks, such as RELi or Green Star

These frameworks impose specific metrics, evidence thresholds, and post-completion evaluations. The Building Surveyor ensures that all QA activities align with these frameworks by:

- Cross-referencing documentation against required credits or compliance matrices
- Tracking the installation and commissioning of systems that impact key criteria (e.g., air quality, lighting, energy)
- Ensuring that submittals and commissioning documentation are submitted, reviewed, and signed off prior to practical completion

Where digital tools are employed (e.g., CDE platforms, BIM-linked commissioning logs, or QR-coded O&M manuals), Building Surveyors may be responsible for integrating QA records into these systems, ensuring smooth handover and future facilities integration.

Proactive Detection of Deviations and Preventing Rework

Construction errors, misalignments, or poor integration between systems can erode performance significantly, and rectifying these issues post-handover is often costly and disruptive. Building Surveyors use their QA role to detect such deviations early, mitigating risk and protecting the project's viability.

Common examples include:

- Mismatched duct sizes or configurations that reduce HVAC efficiency
- Insulation discontinuities that compromise thermal performance and create condensation risks
- Incorrect firestopping details that invalidate compartmentation ratings
- Sensor misplacement or mislabelling that impairs BMS functionality

Through pre-emptive inspections, scheduled QA checkpoints, and non-conformance reporting (NCRs), Building Surveyors flag issues before they escalate—often saving tens of thousands of pounds in rework and avoiding post-occupancy underperformance that could jeopardize warranties, certifications, or end-user satisfaction.

Conclusion: Building Surveyors as Champions of Performance Integrity

In PBBD, the Quality Assurance process is not ancillary, it is the operational core of delivering performance integrity. Through meticulous planning, technical understanding, and hands-on oversight, Building Surveyors ensure that each component of a performance-based system not only meets specifications, but functions seamlessly within the wider building ecosystem.

They are the custodians of precision, the guardians of design intent, and the enablers of long-term building success. By safeguarding quality throughout construction, Building Surveyors help ensure that performance is not only promised, but achieved, validated, and sustained.

Compliance Management and Regulatory Navigation

In traditional design and construction approaches, regulatory compliance typically involves adherence to prescriptive codes and standards, checklists of predefined solutions deemed acceptable by building authorities. However, in performance-based building design (PBBD), compliance is fundamentally reinterpreted. It becomes a matter of demonstrating that building elements and systems achieve or surpass performance benchmarks, even if they depart from established norms.

This shift from rule-following to outcome-based justification introduces a new level of complexity, responsibility, and strategic negotiation, and it is here that the Building Surveyor plays a pivotal role. As technical professionals with a deep understanding of regulatory frameworks, performance criteria, and construction methodologies, Building Surveyors serve as both interpreters and enforcers of compliance in performance-led projects.

They ensure that innovation does not come at the cost of safety, legality, or accountability, while also helping design teams to navigate a more flexible, but potentially more ambiguous, regulatory landscape.

From Prescriptive Codes to Performance-Based Justification

Performance-based compliance allows design teams to pursue alternative methods, materials, and system configurations that depart from traditional solutions, provided that they can prove their effectiveness through technical substantiation. This model is essential for supporting:

- Innovation in sustainability (e.g. passive systems, net-zero energy designs)
- Customised fire strategies for complex geometries or mixed-use spaces
- Adaptive reuse projects with non-conforming existing conditions
- The adoption of smart or modular construction methods

However, this flexibility comes with significant documentation, justification, and approval requirements. Building Surveyors play a crucial role in coordinating these efforts, ensuring that each performance claim is technically valid, regulatorily acceptable, and verifiable in execution.

Core Responsibilities of the Building Surveyor in Regulatory Navigation

Interpreting the Application of Building Regulations

In countries such as the UK, building regulations (e.g., the Approved Documents) outline minimum standards across a range of domains, fire safety, energy efficiency, structural integrity, accessibility, etc. While these documents provide prescriptive paths to compliance, they also allow for alternative approaches, provided that the functional requirements are met.

Building Surveyors assess:

- Whether a performance-based solution (e.g., natural ventilation in lieu of mechanical extract) meets the intent of Part F (Ventilation)
- If a proposed fire strategy aligns with the functional objectives of Part B, even without standard escape stairs or travel distances
- How a structural innovation (e.g., timber construction in high-rise form) satisfies the requirements of Part A for stability and disproportionate collapse

Their role is to interpret the spirit of the regulation, not just the literal requirements—ensuring that creative, tailored, or technically advanced solutions are both compliant and robust.

Validating Alternative Means and Methods Through Collaboration

Performance-based compliance is not achieved in isolation. Building Surveyors work collaboratively with:

- Fire engineers to justify egress routes, smoke control systems, or structural fire performance through quantitative fire risk assessments, CFD modelling, and test data
- Sustainability consultants to validate low-carbon design solutions that may fall outside prescriptive thermal standards
- Designers and M&E engineers to assess whether the integration of unconventional systems (e.g., passive downdraught cooling, hybrid structural assemblies) aligns with the project's compliance trajectory

In doing so, Building Surveyors act as compliance coordinators, ensuring that all disciplines contribute toward a coherent, defensible performance strategy that satisfies both functional and legal expectations.

Engaging with Statutory Authorities and Securing Approvals

Innovative designs often require proactive and transparent engagement with regulators, including:

- Local Authority Building Control (LABC)
- Approved Inspectors (in jurisdictions where applicable)
- Fire and Rescue Services
- Planning authorities or environmental health departments

Building Surveyors lead or support this engagement by:

- Preparing and submitting compliance reports, deviation logs, and alternative solution statements
- Presenting technical evidence, testing documentation, and performance data to support novel design elements
- Negotiating on points of interpretation or demonstrating equivalence through precedent, research, or expert consultation

They are responsible for building the regulatory case for performance compliance, often facilitating peer reviews or engaging third-party certifiers where necessary.

Enabling Innovation Without Compromising Legal or Safety Obligations

One of the most important contributions of Building Surveyors in PBBD is their ability to support design freedom without sacrificing defensibility. This means ensuring that creative or bespoke solutions:

- Are not just permitted, but documented and demonstrably effective
- Do not introduce unintended risks to life safety, occupant health, or long-term maintenance
- Comply with the underlying goals of legislation, particularly in life-critical domains such as fire safety, structural robustness, and means of escape

For example:

- In a high-rise timber office building, a Building Surveyor works with fire engineers to justify the use of mass timber through encapsulation strategies, full-scale fire testing, and comparative risk analysis, ensuring that the structure meets (or exceeds) fire resistance benchmarks.
- In a net-zero education building, a Building Surveyor supports a mixed-mode ventilation design that forgoes traditional mechanical ventilation by demonstrating through simulation and sensor data that indoor air quality, thermal comfort, and humidity control will meet statutory thresholds.

Conclusion: Building Surveyors as Regulatory Strategists

In the performance-based design environment, compliance is no longer a box to be ticked, it is a narrative to be built, supported, and validated. Building Surveyors are uniquely positioned to manage this process, bringing together a deep understanding of construction realities, legal frameworks, and technical performance criteria.

They are not simply regulators, but facilitators of responsible innovation. Their work ensures that forward-thinking design remains grounded in safety, legality, and trust. By navigating the regulatory terrain with precision and pragmatism, Building Surveyors enable design teams to explore new forms, methods, and technologies, secure in the knowledge that compliance is not only achieved, but resilient under scrutiny.

Lifecycle Planning and Maintenance Alignment

In performance-based building design (PBBD), achieving performance targets at the point of handover is no longer sufficient. Today's buildings are expected to function effectively and efficiently over extended lifespans, often spanning 40, 60, or even 100 years. As such, a shift in focus is required: from short-term delivery to long-term operational resilience and from construction completion to lifecycle continuity.

This shift brings the Building Surveyor into a new strategic role, not merely as a construction-phase observer, but as a lifecycle advisor responsible for aligning design intent, system durability, and maintenance strategy from inception through to occupation and beyond.

Building Surveyors are increasingly recognised as essential contributors to whole-life planning, where performance, maintenance, and cost are holistically integrated to ensure that buildings remain safe, compliant, functional, and economically sustainable over time.

The Lifecycle Imperative in PBBD

In a PBBD framework, buildings are designed around specific performance outcomes, thermal efficiency, fire resilience, indoor air quality, energy consumption, acoustic comfort, and so on. However, these outcomes are rarely static; they depend on:

- The ongoing functionality and condition of systems and components
- Timely, appropriate maintenance and replacement
- Monitoring, testing, and recalibration of performance-critical equipment
- The building's ability to adapt to changing use, climate, or regulation

This makes lifecycle planning a fundamental part of performance assurance—and Building Surveyors are uniquely placed to ensure that the performance delivered at practical completion is also the performance maintained across decades of occupation.

Key Roles of Building Surveyors in Lifecycle Planning

Material and System Selection for Maintainability and Longevity

During the design development phase, Building Surveyors contribute insight into the durability, accessibility, and serviceability of proposed materials and systems. Their recommendations go beyond aesthetic or initial performance properties and focus on:

- Expected wear cycles, resistance to environmental degradation, and ease of repair
- Access requirements for maintenance or replacement (e.g. rooftop plant, façade cleaning, concealed ductwork)
- Compatibility with maintenance regimes (e.g. whether a façade system needs bi-annual cleaning or specialist equipment)

For instance, when reviewing a proposed natural ventilation strategy, the Building Surveyor may flag that actuator motors for high-level windows should be located in accessible ceiling voids rather than behind fixed panels, reducing long-term access costs and minimising operational disruption.

Development of Maintenance Schedules and Asset Registers

A core responsibility of Building Surveyors is to ensure that buildings are handed over with clear, detailed, and performance-aligned maintenance strategies. This includes:

- **Planned Preventative Maintenance (PPM) schedules:** defining the frequency and method of inspection, servicing, and recalibration for all performance-critical systems (e.g., smoke detectors, solar PV inverters, HVAC sensors)
- **Asset registers:** itemised inventories of plant, systems, and assemblies with warranty periods, replacement intervals, and lifecycle expectations
- **Lifecycle monitoring strategies:** particularly where predictive maintenance or sensor-based alerts are used to anticipate degradation

These inputs help Facilities Managers understand how the building should be maintained to preserve its intended performance and value.

Alignment with Lifecycle Costing and FM Strategy

Building Surveyors frequently work alongside Quantity Surveyors and Facilities Managers to ensure that maintenance requirements are not only defined but also:

- Budgeted for within lifecycle cost plans
- Phased strategically to avoid disruptive or clustered replacement events
- Integrated into procurement, leasing, and asset management frameworks

For example, in a low-carbon building using a heat pump system, the Building Surveyor ensures that the replacement of inverters and filters is synchronised with manufacturer guidelines, energy performance contracts, and FM capability. They may also advise on spare part provisioning, training needs for in-house FM teams, or outsourcing requirements.

This ensures that long-term operational performance is not compromised by underfunded or misaligned maintenance regimes.

Supporting Digital Handover and Performance Continuity

In advanced PBBD projects, Building Surveyors play an instrumental role in the digital transition from construction to operation. They help develop and verify digital handover packages that include:

- COBie (Construction-Operations Building information exchange) data for all maintainable assets
- Digital twins that link as-built data to real-time performance monitoring tools
- Commissioning records and QA documentation that can be referenced during troubleshooting or refurbishment

This ensures that the building's operational phase is informed by accurate, complete, and usable data, enabling FM teams to take a proactive, rather than reactive, approach to performance management.

Moreover, in smart buildings with extensive sensor networks, Building Surveyors may support the integration of data streams into asset management platforms, allowing for real-time diagnostics, remote monitoring, and machine learning-based predictive maintenance.

Illustrative Example: Passive Design and Lifecycle Planning

In a performance-based passive design project, a school's ventilation strategy relies on automated windows to regulate indoor air quality and temperature. Building Surveyors were involved in:

- Ensuring that the window actuators were accessible for maintenance (e.g., via ceiling hatches rather than scaffold)
- Verifying that actuators were included in the COBie asset list, with correct part numbers, operating instructions, and replacement guidance
- Reviewing O&M manuals to confirm that cleaning and calibration intervals were clearly specified
- Advising the design team to include manual overrides for use during power outages or system faults

Their intervention ensured that the natural ventilation system would not just work at handover, but would remain operational, maintainable, and safe for the life of the building.

Conclusion: Performance Over Time, Not Just on Paper

The involvement of Building Surveyors in lifecycle planning reinforces one of the core principles of PBBD: that performance is not a one-time achievement, but a long-term commitment. Through their work on durability, maintainability, digital handover, and maintenance strategy, Building Surveyors ensure that the designed performance is matched by operational reality, not only during commissioning, but year after year, for decades to come.

Their role is central to creating buildings that are as resilient in operation as they are ambitious in design; buildings that maintain their performance under the pressures of use, weather, wear, and change. In this regard, Building Surveyors act as the guardians of the building's ongoing value, function, and integrity across its full life cycle.

Supporting Post-Occupancy Evaluation and Feedback Loops

One of the defining features of performance-based building design (PBBD) is its commitment to outcomes—not just specifications. This means that a building's success cannot be fully measured at handover. It must be evaluated in use, under real operational conditions, to determine whether it delivers on its performance promises. In this context, Post-Occupancy Evaluation (POE) is not an optional quality add-on but a critical process of validation, learning, and improvement.

Building Surveyors play an increasingly central role in supporting and managing this feedback process. They act as intermediaries between the design team, the facility's users, and operational management, helping to capture, interpret, and respond to real-world performance data. Their involvement strengthens the integrity of performance-based design by ensuring that the loop between design intent and operational outcomes is closed, creating buildings that are not only well-conceived, but also demonstrably effective.

The Evolving Role of Building Surveyors in Post-Occupancy Evaluation

Traditionally, Building Surveyors have been associated with defect identification, maintenance planning, or asset condition surveys conducted years after project completion. In PBBD, their role evolves to include active participation in early-stage POE processes, typically within the first 12–36 months of building operation.

Their responsibilities may include:

- Coordinating POE activities in collaboration with facilities managers, sustainability consultants, and user groups
- Interpreting sensor data, operational records, and BMS outputs in light of the original design targets

- Investigating discrepancies between predicted and actual performance, such as energy use, thermal comfort, or occupant satisfaction
- Recommending technical or behavioural interventions to improve building operation and maintain compliance with performance benchmarks

Through this involvement, Building Surveyors become critical contributors to performance accountability and continuous improvement.

Key Focus Areas in Post-Occupancy Evaluation

Actual Energy Use vs Predicted Consumption

One of the most common gaps in PBBD projects is the discrepancy between energy performance simulation outputs (e.g. SAP, SBEM, or dynamic thermal modelling) and real-world utility bills or submetering data. These discrepancies can result from:

- Differences in user behaviour or occupancy patterns
- Commissioning flaws or system misconfiguration
- BMS overrides or sensor malfunctions
- Design assumptions not materialising as planned

Building Surveyors help collect and contextualise this data, working with energy managers and engineers to identify root causes, quantify their impact, and advise on remedial actions—ranging from retuning controls and sensors to proposing fabric retrofits or additional training for occupants.

Indoor Environmental Quality (IEQ)

Beyond energy metrics, Building Surveyors contribute to evaluating indoor environmental quality, a crucial determinant of occupant health, comfort, and productivity. They may:

- Coordinate or review measurements of temperature, humidity, CO_2 levels, volatile organic compounds (VOCs), and lighting
- Assess whether actual environmental conditions meet design expectations or WELL/BREEAM targets
- Identify physical issues (e.g., poorly sealed windows, unbalanced HVAC, solar glare) affecting user comfort
- Recommend recommissioning, calibration, or retrofit actions to address environmental shortfalls

These assessments are particularly relevant in educational, healthcare, and commercial environments, where comfort and well-being have demonstrable impacts on performance outcomes.

User Satisfaction and Operational Feedback

A building that meets technical targets but frustrates its users can still be considered a performance failure. Building Surveyors play a key role in interpreting subjective feedback within the broader performance context. They may:

- Support user surveys, interviews, and walkthroughs to gather insight into comfort, usability, accessibility, and aesthetic impressions
- Identify misalignments between user behaviour and system operation (e.g. manual window opening disrupting a controlled ventilation strategy)
- Facilitate feedback sessions with occupants and FM teams to gather practical suggestions and report recurring issues

This qualitative data, when triangulated with technical performance findings, allows for a holistic view of building success, one that values both engineering and experience.

Linking POE to Future Performance and Asset Planning

The role of the Building Surveyor extends beyond evaluation into strategic feedback integration. Their insights feed directly into:

- Refinements to the building's maintenance regime (e.g. adjusting filter replacement intervals based on real IAQ data)
- Updates to lifecycle forecasts and cost models (e.g. earlier-than-expected replacement of certain assets)
- Inputs for future refurbishment or expansion projects
- Design guidance for subsequent buildings in the portfolio, especially for institutional or repeat clients

In large organisations, such as universities, hospital trusts, or housing associations—Building Surveyors often become key knowledge holders who link POE findings from one project to the briefing and design process of the next.

Driving a Culture of Performance Accountability

Perhaps most significantly, Building Surveyors help to embed a culture of transparency and learning within project delivery teams. Their involvement in POE:

- Reinforces the idea that performance is not assumed, it must be measured, interpreted, and improved
- Encourages multidisciplinary collaboration between designers, contractors, FM professionals, and end users

- Supports post-handover service offerings, such as soft landings and performance monitoring contracts, creating continuity between construction and operation

This approach ensures that lessons learned on one project are not lost but instead applied to raise the performance standards of future developments.

Conclusion: From Completion to Continuity

In performance-based building design, completion is not the finish line, it's the beginning of verification. Through post-occupancy evaluation and performance feedback loops, Building Surveyors ensure that buildings not only meet expectations, but continue to evolve and improve through their operational life.

By validating performance outcomes, identifying discrepancies, and informing strategic improvements, Building Surveyors support a closed-loop model of design accountability, a model in which data, dialogue, and diagnostics drive better buildings.

They are not merely assessors of completed work, they are custodians of long-term building performance, helping to turn high-performance aspirations into real, measurable, and enduring results.

Examples of Surveyors Role in Performance Based Building Design

Healthcare Facility with Resilience Objectives – Fire Compartmentation Assurance

Project Context: A new acute-care wing in an urban hospital was designed with stringent fire and resilience performance targets, including 60-minute compartmentation in critical care zones and uninterrupted access routes for emergency responders.

Building Surveyor's Role:

- Conducted inspections during wall and ceiling installation to ensure fire-stopping was continuous, documented, and correctly certified.
- Identified a non-compliant service penetration that breached compartmentation and raised a non-conformance report (NCR).
- Coordinated with fire engineers and the contractor to implement a corrective solution, supported by third-party test evidence and product certification.
- Ensured photographic and QA documentation was compiled into the handover pack to demonstrate compliance with Part B of the UK Building Regulations and the hospital's internal fire policy.

Impact: The Building Surveyor helped prevent a critical compliance failure, avoided costly rework post-completion, and safeguarded patient and staff safety during future emergencies.

High-Performance Educational Building – POE and Performance Optimisation

Project Context: A secondary school was designed to Passivhaus standards (German Low Energy Standards) with a hybrid natural ventilation system using automated windows and thermal mass to regulate internal temperatures.

Building Surveyor's Role:

- Conducted a Post-Occupancy Evaluation (POE) 12 months after handover, comparing predicted energy use with actual consumption and collecting indoor air quality data (CO_2, temperature).
- Identified that classroom CO_2 levels frequently exceeded thresholds due to poor actuator timing and user overrides.
- Facilitated a workshop with the design team, controls supplier, and facilities staff to recalibrate window opening strategies.
- Recommended the addition of user feedback panels and manual override signage to improve usability without compromising performance.

Impact: Improved indoor air quality, reduced reliance on mechanical backup systems, and helped the local authority apply findings to other school projects in its estate.

Net-Zero Commercial Office – Lifecycle Maintenance Planning

Project Context: A multi-tenant office development in central London targeted BREEAM Outstanding and operational net-zero carbon certification.

Building Surveyor's Role:

- Reviewed proposed HVAC and façade access systems for maintainability, working closely with FM providers and QSs to align planned preventative maintenance (PPM) schedules with lifecycle cost modelling.
- Flagged that certain façade cleaning systems required abseil access not supported by permanent anchors, leading to a design modification that integrated monorail access systems.
- Ensured all maintainable plant was captured in COBie-compliant asset registers and linked to building management system (BMS) alert protocols.

- Reviewed O&M documentation for clarity, durability, and accessibility standards, particularly for sensor-based and automated systems.

Impact: Enabled the facilities team to manage the building proactively, extended the useful life of façade and HVAC elements, and supported accurate whole-life cost forecasting.

Mixed-Use Urban Development – Navigating Regulatory Equivalence

Project Context: A high-rise building included retail, residential, and communal areas, and used mass timber construction for much of its superstructure, posing regulatory challenges.

Building Surveyor's Role:

- Collaborated with the fire engineer to justify the timber structure using full-scale fire test data, charring rate calculations, and encapsulation strategies.
- Prepared a compliance matrix mapping alternative solutions against the functional requirements of Part B (Fire Safety) of the Building Regulations.
- Led meetings with the local authority building control officer and fire brigade to negotiate acceptance of the fire strategy, supported by digital modelling outputs and risk assessments.
- Monitored construction of fire-resistant zones and documented evidence for insurance and fire certification bodies.

Impact: Helped secure regulatory approval for a technically innovative project, while ensuring the fire strategy remained legally robust and fully documented.

Refurbishment of a Historic Public Building – Materials Compatibility and Maintenance Alignment

Project Context: The restoration of a listed municipal library aimed to integrate high-performance environmental systems into the existing envelope without compromising heritage features.

Building Surveyor's Role:

- Assessed compatibility of proposed breathable insulation with existing masonry walls to prevent trapped moisture and degradation.
- Advised on service routing to minimise interventions in protected finishes, while enabling access for ongoing maintenance and system updates.

- Worked with conservation officers to identify suitable materials that met thermal performance goals while maintaining listed status.
- Coordinated updates to the building's asset register, incorporating new energy systems and tracking heritage-sensitive interventions.

Impact: Enabled a successful blend of heritage conservation and contemporary performance enhancement, reducing the risk of maintenance conflict or regulatory breach.

Summary Table of Surveyor Roles by Project Type

Project Type	Surveyor Role	Key Outcome
Acute Hospital	Fire compliance inspections and QA	Improved life safety and regulatory compliance
Passivhaus School	POE and system recalibration	Enhanced air quality and performance verification
Net-Zero Office	Lifecycle asset planning and digital handover	Improved FM efficiency and cost forecasting
High-Rise Timber Mixed-Use	Performance-based fire compliance documentation	Regulatory approval for non-standard design
Listed Library Refurbishment	Materials selection and access strategy	Balanced conservation and performance upgrades

Conclusion: Building Surveyors as Guardians of Performance Integrity

In the era of performance-based building design, where success is no longer measured solely by initial compliance or design ambition but by how buildings function, adapt, and endure over time, the role of the Building Surveyor has undergone a profound transformation. No longer limited to periodic inspections or defect identification, Building Surveyors have become strategic enablers of performance, guardians of integrity, and custodians of long-term value.

Their presence throughout the project lifecycle, from conceptual design to post-occupancy evaluation, ensures a continuous thread of technical oversight, regulatory navigation, and performance accountability. They play a pivotal role in translating theoretical performance intent into built reality, and critically, in ensuring that performance is sustained, validated, and improved long after practical completion.

Multifaceted Expertise Across the Lifecycle

The strength of the Building Surveyor in PBBD lies in their cross-disciplinary competence. They operate at the intersection of architecture, engineering, building services, regulation, and asset management. This breadth allows them to contribute meaningfully across a wide array of responsibilities, including:

- **Quality Assurance:** Through the design and enforcement of tailored QA frameworks, Building Surveyors verify that installations, assemblies, and systems are executed to a standard that supports performance outcomes, not just construction tolerances.
- **Regulatory Strategy:** In performance-led projects, where alternative approaches must be justified through evidence rather than prescription, Building Surveyors act as interpreters and advocates—ensuring that innovation remains legally defensible and technically robust.
- **Lifecycle Alignment:** By advising on material durability, maintenance access, and operational interfaces, they ensure that long-term usability and cost-effectiveness are embedded into the fabric of the building from the outset.
- **Post-Occupancy Evaluation (POE):** They support data collection, user engagement, and performance analysis after handover, helping to validate design assumptions and guide strategic improvements.

This integrative role positions the Building Surveyor as a performance steward, whose input not only prevents failure but actively promotes functionality, efficiency, and resilience across the entire building lifecycle.

Bridging the Gap Between Vision and Operation

Performance-based design is inherently ambitious. It requires a building to respond to nuanced metrics, thermal comfort, acoustic privacy, daylight autonomy, energy efficiency and fire resilience, often in environments that are dynamic and user dependent. The Building Surveyor's job is to ensure that these ambitions are not just aspirational, but deliverable and durable.

Their site-based presence provides a crucial feedback loop. While architects envision, engineers' model, and contractors build, the Building Surveyor monitors, tests, and reconciles, ensuring that as-built conditions match design expectations, that systems are correctly integrated, and that documentation reflects reality.

This rigorous oversight builds trust across the project team and among external stakeholders. It demonstrates that performance targets are not speculative—they are quantified, verified, and repeatable. In this way, Building Surveyors serve as anchors of accountability, reinforcing the credibility of performance-based design practices.

Elevating the Profession in a Performance-Driven Future

Perhaps most importantly, the expanded role of Building Surveyors in PBBD contributes to the redefinition and elevation of the profession itself. It positions them not as passive overseers, but as active participants in high-performance delivery. It requires new skills, data literacy, systems thinking, digital handover processes, as well as renewed confidence in working collaboratively with architects, engineers, sustainability consultants, and facilities managers.

In rising to this challenge, Building Surveyors are not only supporting better buildings, they are shaping a new model of professional practice: one rooted in foresight, technical precision, and a commitment to long-term outcomes.

Final Reflection: Performance as a Living Commitment

The essence of PBBD is that performance is not a static attribute, it is a living commitment, sustained through use, adaptation, and stewardship. Building Surveyors are the professionals best equipped to uphold that commitment. Through their vigilance, their judgement, and their insistence on aligning design, delivery, and operation, they ensure that buildings are not only built to perform, but continue to perform, for users, for owners, and for the environment.

As guardians of performance integrity, Building Surveyors are central to the future of our built environment, where quality is proven, outcomes are sustained, and buildings truly serve the people and purposes they were designed for.

Chapter 10

Role of Facilities Managers and End Users in Performance-Based Building Design

Introduction: Operational Intelligence in Performance-Based Design

In conventional construction models, Facilities Managers (FMs) and end users are often introduced into the process only after practical completion, once the building is fully designed, constructed, and handed over. At this point, they are expected to operate, manage, and maintain a built environment that they had little opportunity to influence. While this approach may suffice in prescriptive, standardised projects, it undermines the core principles of performance-based building design (PBBD), where the success of a building is judged not by conformance to predefined rules, but by its real-world ability to achieve and sustain specific performance outcomes.

These outcomes, ranging from energy efficiency, thermal comfort, and indoor air quality, to system reliability, user satisfaction, and lifecycle resilience, can only be validated through ongoing use. This makes the experiential, operational, and technical insights of Facilities Managers and end users indispensable to the design process. They are the people who interact with the building day in, day out. They understand the nuances of how it functions under dynamic conditions, how occupants respond to different systems, and where inefficiencies or usability issues typically arise. They also bear responsibility for ensuring that the building continues to perform well throughout its life cycle, well beyond the designer's or contractor's involvement.

In this context, early and continuous engagement of FMs and end users becomes not just valuable, but essential. Their involvement enables design teams to understand:

- How proposed systems will be operated and maintained over time
- Whether the complexity of technologies aligns with available expertise and budget
- How users typically behave in the type of space being designed
- What historical performance gaps or recurring operational challenges exist in similar buildings

Through this dialogue, FMs and users ground design aspirations in operational reality. They help steer designers away from over-specified systems that are difficult to maintain or prone to override, and toward solutions that are practical, intuitive, durable, and cost-effective. Their perspective supports performance-based outcomes by ensuring that technical strategies, such as natural ventilation, sensor-driven lighting, or hybrid energy systems, are not just possible in theory, but sustainable and functional in practice.

Moreover, in the post-occupancy phase, Facilities Managers serve as the continuity agents of performance. They inherit the systems, technologies, and responsibilities for maintaining comfort, safety, and compliance. Their ability to interpret data, identify trends, resolve user issues, and feed observations back into design processes makes them key players in closing the performance loop. Their partnership with designers, surveyors, and energy consultants helps ensure that buildings evolve intelligently, rather than degrading or underperforming over time.

In performance-based design, where buildings must deliver measurable, long-term value, the contributions of FMs and end users are no longer peripheral, they are central to the design ethos. They are the stewards of performance, the first to experience its failures or successes, and the essential link between design intent and lived experience. Their integration throughout the building lifecycle is not merely a best practice, it is a prerequisite for achieving truly high-performing, resilient, and user-centred built environments.

Informing Operational Needs During Design: Embedding Practical Intelligence into Performance Planning

One of the foundational principles of performance-based building design (PBBD) is that a building must be judged by its ability to deliver measurable outcomes during actual use, not merely its compliance on paper or its aesthetic appeal. To achieve this, it is essential that the design process integrates operational intelligence from those who know buildings best: Facilities Managers (FMs).

Facilities Managers bring frontline, experiential knowledge of how buildings function under real-world conditions, knowledge that is often absent from conceptual or simulation-based design processes. Their involvement in early design discussions provides an invaluable operational lens that can help identify design decisions likely to impede usability, maintenance, or energy performance. Their insight ensures that performance is achievable not only at handover, but consistently throughout the building's lifespan.

Operational Insight Areas: The Value of FM Input

Facilities Managers understand, from experience, where buildings succeed or fail in practice. Their contributions cover a broad spectrum of operational concerns:

Spatial Efficiency

FMs consider how building layouts affect circulation, workflow, and equipment movement, especially in high-demand environments like hospitals, data centres, or schools.

- Are corridors wide enough for stretchers or equipment trolleys?
- Can critical spaces be accessed without traversing restricted or sensitive zones?
- Is there adequate space in riser cupboards, plantrooms, or ceiling voids for real-world interventions?

Their input helps avoid congested layouts that may seem functional in drawings but become inefficient or dangerous in practice.

Maintenance Access

Many performance failures arise not from poor design, but from inaccessible systems. If critical systems, such as valves, sensors, dampers, or filters, cannot be reached without removing ceilings, dismantling finishes, or using specialist equipment, maintenance is delayed or neglected, and performance degrades.

- FMs advise on clearances, panel placements, and scaffold-free access.
- They highlight legacy issues from similar buildings and advocate for maintainability as a performance criterion.

Load Management and Zoning

Facilities Managers assess how the building's mechanical, electrical, and plumbing (MEP) systems are likely to respond to changing occupancy patterns or load profiles.

- Are systems zoned appropriately for parts of the building that may be intermittently used (e.g. conference rooms or labs)?
- Are backup or redundancy measures available for mission-critical areas?
- Do HVAC and lighting systems support flexible use and demand-based control?

Their insights help prevent oversizing, underutilisation, or unnecessary energy waste, common culprits of post-occupancy underperformance.

Security and Safety Integration

FMs ensure that access control, alarm systems, fire evacuation paths, and critical infrastructure are designed in line with operational protocols and user behaviour.

- Is CCTV coverage comprehensive but minimally intrusive?
- Are key assets protected from tampering or accidental damage?
- Are emergency shut-offs or override switches intuitive and accessible?

They also ensure that fire escape routes, manual release buttons, or emergency comms are placed where they'll function effectively under duress.

Design Elements Shaped by Operational Input

In a performance-based environment, FM feedback directly influences key architectural and engineering decisions, such as:

Plantroom Locations

FMs may recommend locating plantrooms away from high-use zones or above serviceable areas to minimise acoustic transmission, vibration, and service disruption. They also help determine whether rooftop, basement, or decentralised plant arrangements better support resilience and access.

Sensor and Control Strategy Design

FMs advise on the placement and types of sensors for HVAC, lighting, or BMS integration, helping ensure sensors reflect true conditions (e.g. avoiding placement next to heat sources or openable windows). They also test user interfaces for intuitiveness and operational transparency.

Zoning and Controls for Lighting and HVAC

FMs promote logical zoning strategies that support energy savings without sacrificing comfort. For example, enabling different schedules or override settings in administrative offices vs. technical labs or storage rooms.

Service Riser and Cable Route Coordination

FM insight ensures risers are appropriately sized, future-proofed, and accessible, especially important in adaptive reuse projects where future retrofitting is likely. They also ensure that risers align with FM routines, reducing service route congestion and fire risk.

Asset Monitoring and Data Feedback Loops

Facilities Managers help define which building systems should be monitored, how performance alerts should be prioritised, and which data should be made visible to FM teams. Their input ensures the building management system (BMS) dashboard supports informed decision-making, rather than overwhelming operators with data noise.

Practical Example: Infection Control in a Healthcare Environment

In a hospital design scenario, FM involvement during the planning stage led to the repositioning of service panels and electrical controls for patient isolation rooms. Originally located inside the treatment zone, these panels would have required staff to enter sterile environments for routine maintenance, posing a hygiene risk and operational disruption.

FM input prompted a redesign where access was granted from adjacent service corridors using lockable hatches and sealed chase walls. This:

- Reduced infection control risk
- Simplified maintenance access
- Aligned with performance targets for resilience, operational uptime, and clinical safety

This change, simple on paper, had profound implications for long-term performance and patient care outcomes.

Conclusion: Designing for Reality, Not Just Theory

In PBBD, buildings must perform in the real world, not just in simulations or compliance reports. Facilities Managers, with their operational expertise and institutional knowledge, play a critical role in bridging the gap between idealised design and daily functionality. Their contributions ensure that systems can be accessed, spaces can be navigated, and performance can be sustained over time.

By engaging FMs early in the design process, design teams gain insight into what truly makes a building work, not just at handover, but throughout its operational life. This input leads to buildings that are not only high-performing and efficient, but also usable, resilient, and future-ready, the very essence of performance-based design.

Evaluating Maintainability and Lifecycle Risk: Aligning Technical Ambition with Operational Reality

As buildings become more complex, integrated, and performance-driven, the systems that support their operation, from high-efficiency HVAC units and responsive façades to digital building management systems and renewable technologies, must be not only innovative, but also maintainable and economically sustainable over time. In performance-based building design (PBBD), where long-term functional outcomes are key, evaluating the maintainability and lifecycle risk of proposed systems is a critical component of design success.

Facilities Managers (FMs) bring an irreplaceable dimension to this evaluation. They are the professionals who inherit the building after handover, responsible for keeping its systems functional, its occupants safe and comfortable, and its performance metrics on track. Their institutional knowledge, combined with hands-on experience, allows them to spot future operational challenges that may not be obvious during the design stage.

In PBBD, involving FMs in evaluating maintainability and lifecycle risk helps ensure that design intent aligns with operational capacity, that what is specified can realistically be maintained, repaired, and replaced without compromising building performance or overstretching available resources.

Balancing Advanced Performance with Practical Maintenance

Contemporary performance-based buildings often feature:

- Smart façades that adjust shading or reflectivity in response to light and temperature
- Building management systems (BMS) with occupancy-based control logic
- High-efficiency HVAC equipment using low-global-warming-potential refrigerants and modular VRF systems.
- Renewables such as photovoltaic panels, battery storage, and greywater recycling systems
- Sensor networks and IoT-based monitoring that generate continuous data.

These systems enhance building performance, but they also:

- Introduce new maintenance regimes unfamiliar to some FM teams.
- Depend on specialist subcontractors, spare parts, or digital subscriptions.
- Require periodic recalibration, firmware updates, or performance tuning.
- Pose integration challenges if legacy systems are retrofitted.

Without early-stage FM evaluation, such systems risk becoming overcomplicated, under-maintained, or obsolete, undermining the building's long-term value and functionality.

Facilities Manager Contributions to Maintainability Assessment

Assessing Operational Feasibility

FMs evaluate whether building systems can be effectively operated and maintained using the skills, personnel, and budget available within the organisation. This includes questions such as:

- Are the specified systems compatible with existing FM skillsets?
- Will maintenance require access to specialist training, tools, or service contracts?
- Can minor faults be resolved in-house, or do they require manufacturer intervention?

For example, a BMS requiring remote cloud diagnostics and proprietary software might be unmanageable for an in-house FM team without robust IT support or vendor continuity.

Evaluating Manufacturer and Supply Chain Reliability

Facilities Managers understand the importance of spare parts, warranties, and ongoing technical support. They can:

- Flag when specified components are prone to long lead times, short product lifespans, or supply chain disruption.
- Recommend alternative systems with greater market penetration and longer manufacturer guarantees.
- Identify brands or systems with a history of premature failure, firmware obsolescence, or costly proprietary servicing.

This prevents buildings from becoming locked into inflexible service models or relying on components that are difficult to source or replace.

Highlighting Historical Issues and Lessons Learned

Drawing on their experience with similar facilities, FMs can identify:

- Layouts that resulted in inaccessible plant or overly complicated service zones
- Technologies that were underused or frequently overridden by occupants due to poor usability

- Common failure points in façade systems (e.g. actuator malfunctions, seal failures)
- Trends in unscheduled maintenance, downtime, or excessive operational costs

This institutional knowledge provides a real-world filter on theoretical designs, ensuring that mistakes from past projects are not repeated.

Supporting Lifecycle Cost Planning

In collaboration with Quantity Surveyors and asset planners, Facilities Managers can feed real maintenance and replacement data into:

- Lifecycle costing (LCC) models
- Planned preventative maintenance (PPM) schedules.
- Sinking fund estimations for major system renewal
- Procurement strategies for multi-year service agreements

By contributing data such as actual pump failure intervals, average VRF servicing costs, or access equipment requirements for façade systems, FMs help make cost models more accurate, realistic, and aligned with performance goals.

Design Improvements Informed by FM Lifecycle Input

Facilities Managers can reshape early design decisions by asking critical lifecycle-related questions:

- Can the filter in the MVHR system be replaced without removing ceiling tiles?
- Will a glazing unit with dynamic tinting need specialist access for calibration or replacement?
- Is the green roof accessible for routine maintenance without fall arrest equipment?
- Is the BMS interface user-friendly enough for FM staff to respond quickly to faults and alerts?

Their feedback may lead to:

- Choosing modular components with hot-swappable parts
- Positioning plant in accessible mezzanines rather than in tight ceiling voids
- Specifying longer-life materials even at higher CapEx
- Adding local overrides to automated systems to avoid costly service calls

Example: In a university science building, FM consultation during design led to a switch from custom-built HVAC units to modular rooftop AHUs with field-replaceable components. This change:

- Reduced initial CapEx by 5%
- Cut expected lifetime maintenance costs by 15%.
- Enabled in-house servicing rather than dependence on specialist contractors.

Conclusion: Safeguarding Performance Through Realism and Foresight

The promise of performance-based building design lies in delivering measurable outcomes over time, but this promise can only be fulfilled if systems are designed with their entire lifecycle in mind. Facilities Managers are uniquely positioned to offer grounded, experience-based guidance on what can be maintained, what tends to fail, and what can realistically be sustained over decades of use.

By collaborating with design teams to evaluate maintainability and lifecycle risk, Facilities Managers act as a pragmatic counterbalance to overambition. They help transform high-performance intentions into operationally viable realities, preventing performance gaps and ensuring that investments in advanced systems translate into long-term value, not operational burden.

In this way, FMs are not only maintainers of buildings, but they are also co-authors of their performance journey, safeguarding both economic and environmental outcomes from the earliest stages of design.

Usability and System Interaction: Human-Centred Design as a Performance Catalyst

In performance-based building design (PBBD), much emphasis is placed on the technical performance of systems, efficiency ratings, modelling outputs, and simulation-based forecasts. However, even the most advanced, energy-efficient, or resilient systems can fail to deliver their full potential if the people who interact with them, end users, do not understand, trust, or use them properly.

Whether the occupants are building staff, tenants, patients, students, or visitors, their day-to-day behaviours and decisions directly affect operational outcomes. Systems that rely on human interaction, such as natural ventilation, zoned HVAC, internal shading, or smart lighting—must be designed not just for technical correctness, but for usability, intuitiveness, and behavioural alignment.

Involving end users early in the design process is therefore a critical step in optimising building performance. Their insights help ensure that building systems are not only functional, but also welcoming, intuitive, and likely to be used as intended, supporting the performance objectives that PBBD frameworks strive to achieve.

How End Users Influence Performance Outcomes

Performance-based buildings are increasingly defined by interactive systems that depend on user behaviour. These include:

Manual or Automated Window Controls

In buildings with natural or mixed-mode ventilation, users may be expected to open windows to regulate airflow and temperature. If users don't know when, why, or how to interact with these windows, or if their operation conflicts with automated controls, performance can degrade significantly.

HVAC Zone Overrides

Even in buildings with intelligent zoning and occupancy-based scheduling, users often want the ability to manually adjust temperature settings. If override systems are difficult to access, unclear in function, or too easily manipulated, they can result in increased energy consumption or user frustration.

Lighting Dimmers and Occupancy Sensors

Automated lighting systems are designed for efficiency, but if they switch off too quickly, dim unexpectedly, or offer no manual control, users may disable them entirely. Poorly positioned occupancy sensors or complex interfaces can lead to workarounds that undermine design goals.

Internal Shading and Glare Control

Dynamic blinds or louvres can enhance visual and thermal comfort, but their benefits are often lost when users do not know how to adjust them, or if the controls are poorly located or lack feedback.

Wayfinding and Digital Interfaces

Touchscreen kiosks, room booking systems, or mobile apps that control lighting or HVAC must be accessible, inclusive, and intuitive. Poor interface design can lead to underutilisation or avoidance, weakening the building's digital integration strategy.

Designing for Usability: The Case for Early Engagement

By engaging with end users during the design and prototyping phases, designers and engineers can:

- Test assumptions about user behaviour and expectations.
- Identify cultural or organisational habits that affect system use.

- Validate the location, format, and accessibility of controls.
- Ensure instructions, signage, or digital prompts are clear, concise, and context appropriate.
- Build consensus and ownership of new building systems before occupancy

This user-centred design approach aligns with principles from the soft landing's framework, which advocates for smoother transitions from construction to occupation by preparing users and operators for the building's intended mode of operation.

Behavioural Alignment and Training

In some buildings, especially schools, healthcare environments, or workplaces with shift patterns, the performance of systems depends on consistent and informed use. Engaging end users early allows for:

- Tailored user guides, contextual signage, or tooltips embedded in digital interfaces.
- Training sessions or induction walkthroughs for occupants and FM teams
- Simple visual cues that reinforce sustainable behaviours (e.g. green lights on operable windows, feedback from energy dashboards)

Example: In a net-zero office building, user behaviour was causing operable windows to remain closed, even in mild weather conditions ideal for natural ventilation. Energy use increased as the mechanical cooling system had to compensate. Following engagement with users, the design team added:

- Window-side signage explaining how passive cooling works.
- A green/red indicator light based on indoor/outdoor temperature delta.
- Quick reference cards during onboarding

These interventions significantly improved the frequency of appropriate window use and reduced HVAC loads by an estimated 20% in shoulder months.

Common Usability Pitfalls in Performance-Based Buildings

- Overly complex control panels with minimal labelling or feedback
- Lack of consistency between different interface types (e.g. app vs wall control)
- Systems that conflict (e.g. manual window opening disables mechanical ventilation)
- Controls placed out of reach, in poorly lit areas, or behind furniture.
- Lack of feedback when actions are taken (e.g. no confirmation when blinds are activated)

By consulting users during prototyping and post-occupancy reviews, these issues can be identified and resolved early, preserving the intended performance trajectory.

Conclusion: Bridging the Human-Performance Interface

Buildings are only as intelligent as their users allow them to be. In performance-based design, user interaction is a performance system in its own right, requiring as much design attention and strategic thinking as any HVAC system or façade detail.

By understanding and incorporating user needs, behaviours, and expectations, design teams can bridge the gap between technical potential and actual performance. End users, when engaged early, become partners in sustaining high performance. They are more likely to value, maintain, and responsibly use building systems, helping to preserve environmental goals, operational targets, and comfort standards.

In this way, usability is not a soft concern, it is a critical component of building performance, resilience, and long-term success.

Enhancing Post-Occupancy Performance and Feedback: Sustaining Value Through Operational Intelligence

In performance-based building design (PBBD), the completion of construction marks not the end of the design journey, but the beginning of performance verification in practice. The true test of a building's success lies in how well it performs under real-world conditions, across seasons, usage patterns, and occupancy behaviours. Achieving this requires a structured, collaborative feedback process grounded in real-time data, lived experience, and operational insight.

At the centre of this process are Facilities Managers (FMs), the individuals tasked with operating, maintaining, and adapting the building's systems day after day. Their role in Post-Occupancy Evaluation (POE) is pivotal. They act as both observers and analysts, collecting evidence, interpreting performance trends, and identifying gaps between design intent and operational reality.

By contributing to POE, Facilities Managers close the loop between construction and operation, enabling continuous performance refinement, cost efficiency, and long-term building resilience.

Post-Occupancy Evaluation (POE) in the PBBD Framework

POE is a structured process of evaluating how a building performs once it is in use. It encompasses:

- Quantitative assessments (e.g., energy consumption, CO_2 levels, maintenance logs)

- Qualitative feedback (e.g., user satisfaction, complaints, usability)
- Comparative analysis (e.g., predicted vs actual performance data)

In the PBBD model, POE is not a one-off exercise but part of a cyclical feedback mechanism. It informs adjustments to building systems, fine-tunes occupant interaction strategies, and validates (or challenges) the underlying design assumptions.

Facilities Managers as Performance Interpreters and Performance Guardians

Facilities Managers play several critical roles in enhancing post-occupancy outcomes:

Performance Monitoring and Data Collection

FMs are responsible for tracking a wide array of performance indicators, including:

- Energy use intensity (EUI) via utility data or sub-metering
- Thermal comfort based on space temperature trends and occupant feedback.
- Indoor Air Quality (IAQ) metrics such as CO_2, VOCs, RH, and PM2.5
- System faults, maintenance calls, and downtime logs

This data is often gathered through Building Management Systems (BMS), smart sensors, and work order tracking software. FMs are uniquely equipped to contextualise this data with knowledge of building use patterns, occupancy schedules, and operational anomalies (e.g., seasonal staff shifts, weather events, or usage surges).

Pattern Recognition and Anomaly Detection

FMs develop an intuitive understanding of how a building "should" behave. They can:

- Detect inefficiencies (e.g., HVAC running overnight, constant override of blinds or lights)
- Identify trends (e.g., increased temperature complaints in specific zones)
- Isolate anomalies (e.g., spikes in water use suggesting leaks or system faults)

Their observational insight helps diagnose root causes of underperformance and supports targeted investigation and correction.

Supporting POE Reviews and Performance Dialogue

FMs actively contribute to formal POE reviews by:

- Bringing forward user complaints, system logs, and non-conformance reports
- Interpreting system interaction behaviours (e.g., why users override controls or ignore interfaces)

- Providing data from maintenance routines, asset condition surveys, and work requests
- Collaborating with designers, Building Surveyors, and commissioning agents to analyse findings and recommend solutions.

Their feedback provides real-world clarity to performance data and helps connect technical metrics with human experience.

Driving Corrective Actions and Performance Optimisation

The insights provided by FMs support:

- Recommissioning of systems (e.g., recalibrating controls, adjusting setpoints, or rebalancing HVAC zones)
- Technical interventions (e.g., adding insulation, improving BMS programming, or upgrading filters)
- Occupant training or signage updates to improve user-system interaction.
- Refinement of maintenance protocols based on early wear or failure indicators.

This iterative process enables buildings to adapt intelligently over time, retaining their performance edge rather than gradually drifting away from design intent.

Portfolio-Wide Learning and Organisational Intelligence

In organisations that manage multiple buildings—such as universities, healthcare trusts, retail chains, or corporate estates, Facilities Managers play a strategic role in aggregating and applying lessons learned across projects. They:

- Compare POE findings across buildings with similar systems or layouts.
- Maintain institutional knowledge on recurring issues, best practices, and technology performance.
- Guide procurement and system selection in future projects.
- Inform design briefs, PPM frameworks, and contractor evaluation criteria.

This cross-project feedback enhances organisational learning and supports the development of performance-informed design standards, improving outcomes at scale.

Case in Point: Learning from POE in a School Campus

A Facilities Manager overseeing a school campus noticed that in three newly built classrooms, IAQ complaints were rising despite the presence of a demand-controlled ventilation (DCV) system. POE data revealed:

- CO_2 levels exceeding 1500 ppm in afternoon sessions

- Poor actuator response in operable windows
- Teacher overrides of automated blinds leading to overheating

Through collaboration with the design team and system integrators:

- Actuators were recalibrated for faster response.
- CO_2 sensor placement was improved.
- A staff induction programme was introduced to reinforce passive comfort strategies.

Follow-up POE six months later showed a 35% reduction in high CO_2 episodes, improved occupant satisfaction, and reduced mechanical cooling demand.

Conclusion: Making Performance Endure

Performance-based design does not guarantee performance, it simply sets the stage for it. The ongoing achievement of building performance depends on careful monitoring, responsive action, and structured feedback, led by those closest to the building's operation.

Facilities Managers are not just system operators, they are data analysts, risk mitigators, and collaborators in performance stewardship. Their role in post-occupancy evaluation ensures that buildings not only meet initial expectations, but also adapt, evolve, and improve.

Through their insights and actions, FMs make performance a living attribute, ensuring that design intent translates into durable, demonstrable, and accountable outcomes.

Supporting Digital Handover and Operational Readiness: Bridging the Gap Between Commissioning and Competence

In performance-based building design (PBBD), buildings are increasingly delivered as complex digital-physical hybrids, where performance is monitored, maintained, and optimised through sensor networks, cloud-integrated building management systems (BMS), and rich datasets that document every aspect of construction and operation. The traditional "handover" process, once focused on paper-based manuals and brief orientation sessions, is now a critical transition point requiring robust digital infrastructure, intuitive interfaces, and fully operational facilities teams.

In this context, Facilities Managers (FMs) play a central role in supporting and shaping digital handover and operational readiness. Their early and active involvement ensures that handover data is structured, accessible, and aligned with actual operational needs, enabling a seamless transition from construction to day-to-day performance management.

The Evolution of Building Handover in PBBD

Digital transformation has redefined what it means to "handover" a building. No longer limited to a stack of as-built drawings and O&M manuals, the digital handover now includes:

- COBie (Construction-Operations Building information exchange) datasets.
- Digital twins with embedded maintenance schedules and performance simulations
- Cloud-based maintenance portals with dynamic asset lists and work order tracking
- Mobile interfaces for BMS control and condition monitoring
- Integration with Enterprise Asset Management (EAM) systems or Computerised Maintenance Management Systems (CMMS)

For these tools to be effective, they must be configured with FM workflows in mind, not just technical completeness. Facilities Managers ensure that the digital handover is practical, relevant, and usable, avoiding information overload and data bloat while preserving long-term value.

Key FM Contributions to Digital Handover and Operational Readiness

Participating in the Development of Digital Handover Packages

FMs collaborate with contractors, BIM coordinators, and digital consultants to ensure that digital deliverables are:

- Aligned with actual FM practices and asset management software.
- Structured logically (e.g., zone-based vs system-based grouping)
- Populated with accurate, verified, and use-ready data (e.g., warranty dates, replacement intervals, manufacturer contacts)
- Delivered in accessible formats (e.g., COBie spreadsheets, IFC files, APIs for system integration)

FMs may also specify how data is to be used post-handover, for example, by requesting QR-coded asset tags linked to real-time condition data, or by shaping FM dashboards that prioritise actionable alerts over passive monitoring.

Reviewing O&M Manuals for Clarity and Usefulness

In high-performance buildings, the complexity of system operation often outpaces the capacity of standard handover documentation. Facilities Managers:

- Review manuals to ensure they include step-by-step procedures, clear diagrams, and plain-language troubleshooting guides.
- Verify that documents match as-installed equipment and reflect field conditions.
- Flag omissions, inconsistencies, or excessive technical jargon that would hinder operational readiness.

Effective O&M manuals should serve not just as compliance artefacts, but as working documents for staff training and daily use.

Testing Maintenance Portals and Mobile Interfaces

Modern buildings often feature:

- Web portals for scheduling maintenance and accessing service histories.
- Mobile apps for scanning asset tags, recording faults, or adjusting system settings.
- BMS dashboards accessible remotely or via tablets

FMs test these tools during commissioning to ensure:

- Intuitive navigation and logical menu structures
- Functional compatibility with existing FM software and workflows
- Alert configurations and permission settings that support accountability and response protocols.

They also provide feedback to developers and integrators to tailor interfaces for actual FM use, not just system engineer preferences.

Establishing Alert Thresholds and Maintenance Triggers

Performance-based buildings rely on real-time monitoring and predictive maintenance. FMs work with BMS engineers and commissioning agents to:

- Define alert thresholds for temperature, humidity, airflow, CO_2 levels, and system efficiency.
- Set priority levels and escalation protocols for fault detection.
- Create automated triggers for maintenance actions based on runtime hours, sensor degradation, or operational anomalies.

By configuring these settings with FM input, the building can move beyond reactive fault detection and toward proactive performance management.

Benefits of FM Involvement in Digital Handover

- Reduces onboarding time for FM teams and third-party service providers.
- Minimises operational downtime and confusion post-handover.
- Ensures data integrity, accessibility, and integration with live systems.
- Supports compliance tracking, especially in safety-critical or ESG-driven buildings.
- Enables better preventative maintenance planning, reducing lifecycle costs and enhancing resilience.

Illustrative Example: Digital Twin Implementation in a Commercial Tower

In a smart commercial office tower targeting WELL and LEED Platinum certification, the Facilities Manager was involved from Stage 4 design through to commissioning. During digital handover, the FM:

- Reviewed and modified the COBie dataset structure to align with the organisation's EAM platform.
- Helped configure a digital twin interface to display actionable data such as air quality trends, energy performance, and asset condition states.
- Requested a custom fault escalation protocol in the BMS based on location, urgency, and time of day.
- Co-developed a training module for FM staff embedded within the digital twin, ensuring continuous learning and process alignment.

The result was a fully operational digital handover, reducing post-occupancy service delays and enabling seamless facilities integration from day one.

Conclusion: Readiness, Ownership, and Performance Continuity

In PBBD, the transition from construction to operation is not a moment, it is a strategic phase requiring data clarity, user preparedness, and system usability. Facilities Managers are central to ensuring that this transition is intentional, supported, and successful.

Through early involvement in digital handover processes, FMs become co-authors of operational readiness. They transform technical documentation into operational strategy, align digital systems with real workflows, and enable building teams to take ownership of performance from the outset.

In doing so, they help ensure that performance is not only designed, but delivered, maintained, and improved over time.

Conclusion: Performance is a Partnership

In the context of Performance-Based Building Design (PBBD), the delivery of a building is not the culmination of the design process, it is merely the beginning of its performance journey. Unlike traditional construction models where success is often measured by timely completion or compliance with prescriptive codes, PBBD shifts the focus to long-term functionality, adaptability, and efficiency. In this new paradigm, performance is not a fixed attribute, it is a living commitment that must be upheld, monitored, and refined over time.

Achieving this level of sustained performance requires more than intelligent design and high-specification systems. It requires **collaboration**. It requires the active participation of those who **operate, occupy, and live within the building**. In particular, **Facilities Managers (FMs)** and **end users** are essential partners in translating design vision into real-world success. Their roles are no longer supplementary—they are **central to the performance ecosystem**.

From Design Intent to Operational Reality

Facilities Managers and end users bring ground-level insight and operational realism into what is often a highly theoretical or model-driven design process. They are the individuals who will test the assumptions embedded in simulations, challenge the practicality of maintenance strategies, and reveal how systems are truly used, not just how they were intended to be used.

By engaging these stakeholders from the early design stages, PBBD becomes more than a technical blueprint, it becomes a conversation across disciplines and timeframes. The FM highlights whether plant access is feasible, or if controls are serviceable without downtime. The end user explains how interface complexity might lead to confusion or disengagement. Together, they bridge the gap between prediction and performance, ensuring that what is drawn in the studio survives the test of real-world use.

Operational Insight as a Catalyst for Continuous Improvement

The dynamic nature of building use means that performance is never truly "finished." Occupancy patterns evolve, technologies advance, and environmental conditions shift. The feedback loops created by FMs and users, through post-occupancy evaluation, real-time monitoring, and operational troubleshooting, are vital mechanisms for detecting issues, fine-tuning systems, and adapting strategies.

Their lived experience, when harnessed, feeds directly into:

- Lifecycle cost management
- System recalibration and recommissioning
- Refinement of comfort and usability interfaces
- Informed design of future projects or refurbishments

This continuous feedback supports a culture of learning and resilience, a hallmark of mature performance-based design practices.

Creating Buildings That Perform Today and Evolve for Tomorrow

Buildings are long-lived assets, often expected to perform over decades, through occupancy changes, economic shifts, and climate transitions. Ensuring that they continue to serve their users, support their functions, and meet environmental goals is not solely the domain of the architect or engineer. It is a shared responsibility, where design teams, operations staff and building occupants each play a crucial role.

When FMs and end users are treated as partners rather than afterthoughts, buildings become:

- Easier to operate
- Cheaper to maintain
- More energy-efficient
- More comfortable and usable
- More adaptable to future needs

This human-centric, operations-focused integration transforms PBBD from a static goal into a resilient and adaptive strategy for delivering real-world value.

Final Reflection: A Collaborative Model for Lasting Impact

At its core, performance is a partnership, a collaboration that spans disciplines, project phases, and building lifecycles. Architects, engineers, surveyors, contractors, Facilities Managers, and users must work together not only to create buildings that meet performance standards on day one, but to ensure those standards are maintained, validated, and evolved over time.

This is the essence of PBBD: not a one-time achievement, but a living process of continuous alignment between design aspiration and operational reality. In recognising the critical contributions of those who use and maintain our buildings, we unlock the true potential of high-performance design, not just in metrics, but in meaningful, measurable, and enduring impact.

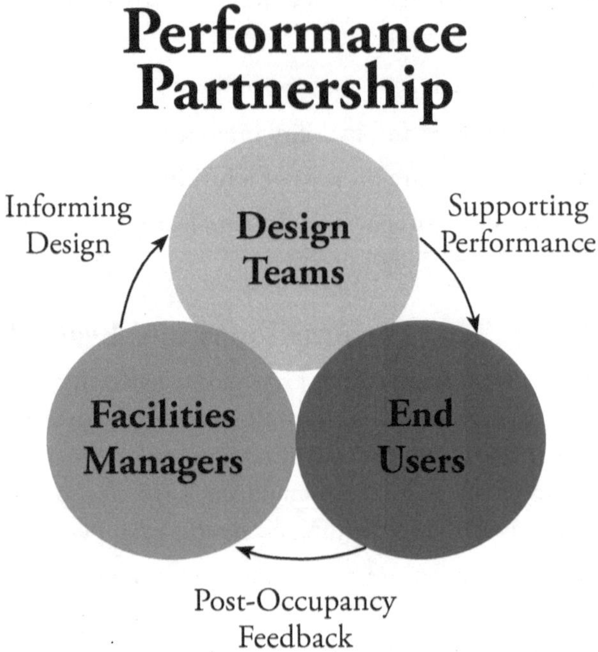

Chapter 11

Role of Building Control in Performance-Based Building Design

Introduction: Building Control in the Era of Performance-Based Design

In the evolving landscape of the built environment, Performance-Based Building Design (PBBD) represents a paradigm shift, one that prioritises measurable outcomes over prescriptive compliance. Rather than conforming to fixed construction methods or standardised assemblies, PBBD empowers designers to achieve regulatory objectives through innovative, context-specific, and evidence-led approaches. In this model, the benchmark for success is not whether the design follows a checklist, but whether it demonstrably delivers on critical functions such as fire safety, structural stability, energy efficiency, accessibility, and occupant well-being.

This shift introduces both opportunity and complexity. While PBBD unlocks greater flexibility and creative freedom, it also necessitates a more sophisticated and interpretive approach to regulatory compliance, one that accounts for equivalency, proportionality, and risk-based assessment. At the heart of this transformation are Building Control bodies, whose role has evolved from code enforcers to strategic validators of performance.

Building Control professionals, whether acting as Local Authority Building Control (LABC) officers or as Approved Inspectors in jurisdictions where private approval services are authorised—serve as the regulatory backbone of performance-led projects. They are tasked with ensuring that buildings are not only designed to function, but are also constructed, verified, and documented in accordance with the functional requirements of national building regulations.

From Prescriptive Oversight to Performance Interpretation

Traditionally, the function of Building Control revolved around reviewing design proposals, inspecting construction progress, and verifying that completed buildings aligned with prescriptive regulatory standards. In PBBD, their role expands considerably. Building Control must now evaluate how non-standard, or novel design strategies can be shown to meet or exceed the intent of the law, often using performance data, modelling outputs, material testing, and third-party certification as evidence.

Their remit extends beyond simple rule enforcement to include:

- **Performance justification:** Reviewing how simulation results or engineering analyses demonstrate compliance with functional objectives (e.g., smoke clearance in lieu of an additional escape stair).
- **Equivalence assessment:** Determining whether an alternative solution, such as a cross-laminated timber (CLT) frame treated for fire performance, achieves a level of safety equivalent to that offered by traditional concrete or steel construction.
- **Legal and procedural accountability:** Ensuring that the project team's use of performance-based methodologies is robust, evidence-backed, and legally defensible in the event of dispute or failure.

In this capacity, Building Control becomes a mediator between innovation and legal responsibility, helping to enable creativity while protecting life safety and structural integrity.

Gatekeepers of Safety, Structure, and Public Interest

One of the core responsibilities of Building Control remains unchanged: to safeguard the public by ensuring that buildings are safe, stable, and fit for purpose. However, in PBBD, achieving these outcomes demands a far more analytical and evidence-driven approach. The performance lens requires Building Control to engage with:

- Alternative fire strategies, where travel distances exceed standard allowances but are mitigated by enhanced alarm systems or compartmentation.
- Complex building geometries or hybrid uses, which defy typical design templates and require nuanced interpretation of structural and access regulations.
- Innovative technologies, such as green roofs, living walls, or smart façades, whose long-term compliance with weathering, durability, and loading criteria must be validated.

Their engagement ensures that design flexibility does not equate to regulatory ambiguity, and that every decision made in pursuit of performance is anchored to statutory objectives, from fire escape times and acoustic separation to thermal resistance and disabled access.

An Evolving Role in a Changing Industry

The rise of PBBD, digital construction tools, and performance-led procurement models is reshaping how buildings are conceived, built, and regulated. Building Control professionals must now be conversant with:

- BIM-integrated workflows
- Digital compliance documentation and 3D visualisations
- Building performance simulations (e.g., CFD, energy modelling)
- Third-party validation schemes and material certifications

This demands a higher level of technical fluency and collaborative engagement with design teams than ever before. Building Control officers are now regular participants in early-stage design reviews, fire engineering strategy meetings, and cross-disciplinary workshops, providing regulatory insight that is proactive rather than reactive.

Their role is not to hinder innovation, but to guide and validate it, ensuring that emerging building typologies and construction methods remain within the bounds of legal and functional safety.

The Regulatory Anchor of Performance Design

As PBBD becomes increasingly central to the delivery of high-performance, sustainable, and adaptable buildings, the role of Building Control is more vital than ever. These professionals serve as the regulatory anchor of the performance-based design process, ensuring that freedom of expression and technical ingenuity do not compromise safety, structural integrity, or public interest.

By evaluating performance-based strategies through the lens of legal equivalence and functional adequacy, Building Control enables the built environment to evolve responsibly, safely, and innovatively, anchoring bold design in a framework of compliance and trust.

Regulatory Interpretation and Performance Equivalence: Evaluating Innovation Through a Legal Lens

In the framework of Performance-Based Building Design (PBBD), regulatory compliance is no longer a binary outcome defined by adherence to pre-written codes. Instead, it becomes a process of interpretation, justification, and performance equivalence, whereby the key question is not "Does this follow the standard guidance?" but "Does this solution achieve the intended level of safety, stability, accessibility, or environmental performance?" This subtle but important shift in mindset places

Building Control professionals at the heart of design assessment, not as enforcers of prescriptive rules, but as technical adjudicators of functional outcomes.

Their primary responsibility is to ensure that all aspects of a building's design align with the statutory requirements set out in frameworks such as the UK Building Regulations 2010, which are accompanied by Approved Documents (Parts A through S). These documents offer prescriptive paths to compliance but also allow for alternative approaches, provided the design team can demonstrate that these alternatives meet or exceed the underlying functional requirements.

Functional Requirements vs Prescriptive Solutions

Each Part of the Building Regulations begins with a set of functional requirements, the overarching safety, health, and welfare goals that buildings must achieve. The Approved Documents then offer one way of meeting those goals, usually based on widely accepted construction methods and risk profiles. However, PBBD embraces the idea that alternative strategies may deliver the same, or even better outcomes, as long as they can be validated through credible performance evidence.

This is where Building Control's interpretive role becomes critical. Officers must assess whether non-standard design proposals:

- Address the core intent of the regulations
- Are supported by credible, verifiable data
- Represent an acceptable level of risk and resilience
- Are practical to implement and enforce on-site

In this way, Building Control provides a structured, legally defensible pathway for design flexibility, innovation, and engineering judgment.

Examples of Performance Equivalence in Practice

Fire Safety Under Part B

A design team proposes a residential tower with a single escape stair, enhanced by a high-integrity pressurised smoke control system and early detection. Under prescriptive guidance, multiple staircases may be required. However, with a performance justification in place, including CFD modelling, evacuation simulations, and test data, the proposal may be deemed to meet the functional requirement for safe means of escape.

Building Control must assess:

- Is the proposed strategy supported by reliable simulation outcomes?

- Are there suitable redundancy and fail-safes in the system?
- Does the occupancy profile support the risk mitigation strategy?

Structural Innovation Under Part A

A university research centre incorporates a hybrid timber-steel structural frame, which departs from typical construction types listed in guidance. To justify compliance, the design team submits:

- Load testing data for the hybrid connections
- Structural performance calculations from an accredited engineer
- Third-party certification and industry precedents

Building Control evaluates whether the proposal offers equivalent load-bearing capacity, deflection control, and robustness compared to conventional options, and whether it aligns with the overarching requirement for structural stability under both normal and exceptional conditions.

Evidence Review and Evaluation Process

When alternative compliance routes are proposed, Building Control must carry out a methodical and evidence-based review, typically involving:

- Engineering and technical reports that model the proposed system's behaviour under relevant load or stress conditions
- Fire safety assessments, often using computational fluid dynamics (CFD) models or egress time simulations
- Material certification, including documentation from UKAS-accredited labs, BBA Agrément Certificates, or CE/UKCA marking
- Risk assessments that compare the proposed solution with baseline risks identified in standard methods
- Precedent studies, where similar systems have been approved or tested successfully on other projects

The process may also include consultation with external stakeholders such as:

- Fire and rescue services
- Planning departments
- Structural peer reviewers
- Insurers and warranty providers

This multi-layered review ensures that no design deviation is accepted lightly, and that all performance-based strategies are rooted in technical rigor, not speculation or cost-saving.

Proportionality and Contextual Judgement

A crucial aspect of Building Control's interpretive role is proportionality. The acceptability of a performance-based approach is rarely universal, it depends on the building's:

- Size and complexity
- Occupancy type and density
- Criticality of use (e.g. schools, hospitals, public infrastructure)
- Fire load, exposure, or environmental risk

For example, a performance-based ventilation strategy may be acceptable in a small office block but require far more substantiation in a care home or high-rise apartment building.

Building Control officers must weigh technical adequacy against contextual risk, ensuring that performance-based solutions are not only theoretically sound, but appropriate to the building's function and user profile.

A Role of Technical Mediation and Public Trust

In PBBD, regulatory interpretation is not about bending rules, it is about upholding them through informed, flexible, and evidence-led decision-making. Building Control officers act as technical mediators who balance innovation with public trust, ensuring that alternative strategies do not diminish safety, usability, or legal accountability.

Their capacity to assess performance equivalence is central to making PBBD a viable, scalable, and responsible design methodology. Without their interpretive oversight, the risk of unregulated experimentation would increase. With it, buildings can evolve safely, thoughtfully, and in service of better performance outcomes for all.

Review and Approval of Performance-Based Strategies: Enabling Innovation Through Regulatory Confidence

A defining strength of Performance-Based Building Design (PBBD) is its flexibility, the freedom it gives design teams to meet regulatory goals using non-prescriptive, performance-led solutions. However, this flexibility must be tempered by oversight, evidence, and public accountability. The authority to confirm that such alternative strategies are appropriate, effective, and legally compliant rests with Building Control.

In this evolving regulatory environment, one of Building Control's most critical responsibilities is the review and approval of performance-based strategies. This process goes well beyond checking boxes against a checklist—it involves a deep understanding of building systems, risk, occupant needs, and the interplay between innovation and legal responsibility. It is Building Control's considered judgment that allows design teams to explore new methods and materials without undermining safety, accessibility, or statutory compliance.

Validating Engineered Alternatives to Prescriptive Solutions

Performance-based approaches often challenge conventional compliance routes. Rather than relying on tabulated values, deemed-to-satisfy provisions, or standardised designs, the PBBD process may involve:

- **Custom fire engineering:** Such as escape time calculations, CFD (computational fluid dynamics) modelling, and smoke control design to support variations from standard egress routes.
- **Energy and environmental simulations:** Used to justify building envelope designs, passive ventilation strategies, or renewable system integrations not explicitly covered in Approved Document Part L.
- **Innovative structural systems:** Including hybrid timber-steel frames, modular components, or advanced composite materials that require full-scale testing and performance modelling to demonstrate structural adequacy under Part A.

Building Control officers are responsible for evaluating whether these solutions meet the intended functional outcomes of the regulations and whether they do so in a way that is technically robust and practically achievable.

Their review may include:

- Scrutinising fire strategies developed by chartered fire engineers, including all relevant modelling assumptions and mitigation measures.
- Examining energy models and dynamic simulations, ensuring consistency with regulatory baselines.
- Verifying that load paths, failure tolerances, and system integration are clearly understood and coordinated across the project's design and construction disciplines.

Approving Emerging Technologies and Non-Standard Products

The increasing demand for low-carbon, efficient, and resilient buildings has driven a wave of innovation in construction materials and systems. PBBD allows for their use, but only when properly assessed.

Building Control evaluates and approves:

- New or untested products supported by Agrément certificates (e.g., BBA), European Technical Assessments (ETAs), or third-party conformity reports.
- Proprietary assemblies and prefabricated systems with no direct comparison in regulatory guidance, provided their performance is validated through testing or certification.
- MEP innovations such as smart systems, adaptive lighting, and high-performance HVAC modules that achieve compliance through modelling and dynamic control logic, not fixed prescriptive performance.

This approval process ensures that novelty does not compromise accountability. It gives confidence to developers, insurers, and end users that unfamiliar solutions have been objectively reviewed and benchmarked against equivalent standards of safety and performance.

Collaborative Regulatory Engagement: A Dialogue, Not a Verdict

Unlike traditional enforcement, the review of performance-based strategies is inherently collaborative. Building Control professionals do not operate in isolation, they function as part of a multi-disciplinary approval ecosystem, engaging directly with:

- Fire engineers, to verify compensatory measures and system reliability under fire conditions.
- Structural engineers, to confirm system behaviour under live and dead loads, disproportionate collapse, and seismic or wind events.
- MEP consultants, to validate simulations and understand system redundancies or control strategies.
- Project architects and contractors, to ensure that approved designs are buildable, maintainable, and accurately interpreted on site.
- Planning and statutory authorities, especially in projects with conservation, high-risk, or community-sensitive elements.

Through design team meetings, plan reviews, formal submissions, and workshops, Building Control can prompt:

- Revisions to assumptions in simulation models

- Introduction of additional mitigation layers (e.g., alarms, compartmentation, access improvements)
- Refinement of justification documents and compliance narratives

This iterative dialogue is not about opposing design freedom, but about co-creating compliance pathways that are defensible, repeatable, and aligned with public interest.

Establishing Legal Defensibility and Confidence

Perhaps the most powerful contribution of Building Control's review is the legal legitimacy it bestows upon non-standard solutions. A performance-based strategy that has been approved by Building Control becomes a documented, auditable, and enforceable basis for project delivery. This provides:

- Reassurance to clients and stakeholders that innovative designs do not introduce uncontrolled risk.
- Protection to contractors and consultants against claims of non-compliance.
- Confidence to facility operators and insurers that building systems have been rigorously reviewed and meet regulatory thresholds.

In the event of future audits, disputes, or operational failures, this documented approval provides a clear chain of reasoning, supported by technical evidence and professional validation.

Conclusion: Enabling Safe Innovation Through Trusted Oversight

As performance-based design continues to redefine the boundaries of what buildings can achieve, the role of Building Control as reviewer and approver becomes increasingly strategic. It is their professional judgment that enables design freedom to flourish, so long as that freedom is backed by science, engineering, and operational clarity.

By engaging deeply with alternative strategies, validating evidence, and participating in collaborative dialogue, Building Control does not simply grant permission, they provide the foundations of trust upon which innovation can safely stand. Their approvals transform bold ideas into constructible, compliant, and high-performing realities, making them indispensable stewards of the performance-based design process.

Ensuring Life Safety and Structural Integrity: Safeguarding Performance Without Compromise

At the core of every building regulation system, regardless of the design approach, is a commitment to protecting life, property, and public safety. While Performance-Based Building Design (PBBD) introduces welcome flexibility in how compliance is achieved, it does not dilute the ultimate purpose of the regulatory framework. In fact, it raises the bar by demanding that performance goals be demonstrated, not merely assumed.

In this context, Building Control officers remain the final custodians of life safety, structural integrity, and accessibility. Their role ensures that in the pursuit of innovation, creativity, or environmental targets, fundamental safety is never compromised. They provide the regulatory and technical oversight that transforms flexibility into responsibility and ensure that any deviation from prescriptive norms is supported by clear evidence, sound engineering, and practical feasibility.

A Non-Negotiable Duty in a Flexible Framework

PBBD allows for alternative compliance strategies, but the outcomes, safe evacuation, structural resilience, universal access, remain non-negotiable. To that end, Building Control officers must make informed, proportionate, and defensible judgments on:

- Whether a fire protection strategy provides equivalent levels of life protection to standard guidance
- Whether a non-standard structural system has sufficient load capacity, robustness, and redundancy
- Whether escape routes, lifts, entrances, and wayfinding provide equal access and egress for persons with disabilities

Their review process is not focused on methods, but on outcomes. The questions they ask are:

- Can this building withstand the expected loads, including live, dead, wind, and exceptional events?
- Will occupants be able to exit safely in a fire scenario, within an acceptable timeframe and with sufficient redundancy?
- Are vulnerable occupants, such as children, elderly, or mobility-impaired individuals—sufficiently accommodated by the access and emergency strategy?

These questions drive every decision they make, from initial design validation to final sign-off.

Evidence-Based Decision-Making: Rigour in Justification

In PBBD, design teams are required to justify their proposals with quantifiable evidence. Building Control officers must assess whether that evidence:

- Meets an accepted technical standard (e.g., BS 9999, BS EN 1991-1-1)
- Has been generated using industry-accepted modelling tools, such as:
  - CFD (Computational Fluid Dynamics) for smoke and heat movement
  - Evacuation Modelling for egress and compartment behaviour
  - Finite Element Analysis (FEA) for structural response to loads
- Is appropriately verified or peer-reviewed, especially in cases of novel or high-risk strategies
- Includes testing and certification from UKAS-accredited labs, or third-party bodies like BBA or BRE

An evidence-based decision by Building Control is one that can be:

- Clearly traced to authoritative standards or accepted industry methodologies
- Understood and reproduced by other professionals
- Defended in the case of legal inquiry or enforcement

Risk-Proportional Review: Context Shapes Compliance

No two buildings are alike, and Building Control officers must evaluate performance-based strategies through the lens of risk proportionality. That means factoring in:

- **Building height and complexity:** Is it a single-storey school or a high-rise residential tower?
- **Occupancy type and density:** Does the building house sleeping occupants, children, or vulnerable users?
- **Use profile:** Will it operate continuously, have high turnover, or contain hazardous activities?
- **Fire load and exposure:** Are materials and contents contributing to increased fire risk?

For example, a hotel with a single escape route may be acceptable with extensive fire detection, pressurisation systems, and emergency lighting, but the same approach in a care home might fail, due to slower occupant response and need for staff-assisted evacuation.

Building Control must calibrate their expectations and approvals to the specific risk profile of the building, recognising that performance equivalence is not universal, it is context dependent.

Transparency and Traceability: Regulatory Assurance and Public Trust

PBBD places new demands on Building Control in terms of documentation, traceability, and defensibility. Every approval, comment, or requested revision must be:

- Clearly recorded, with reference to performance benchmarks or modelling assumptions
- Justified in writing, especially in the case of significant departures from standard design
- Capable of being reviewed in the future—for example, during a compliance audit, post-incident investigation, or retrofit scenario

To support this, many Building Control departments are adopting:

- Digital compliance platforms, linking design submissions with modelling outputs, simulations, and QA documentation
- Risk registers, tracking unresolved issues, mitigation measures, and residual risks
- Audit trails, showing how each decision was arrived at, including communication with design teams and specialists

This transparency fosters accountability, consistency, and legal robustness, ensuring that life safety decisions are not arbitrary, but grounded in best practice.

Illustrative Example: Scrutiny in Fire Strategy Validation

A mixed-use development proposes an open atrium linking multiple floors, with extended travel distances in the residential corridors. The fire engineer submits a strategy using:

- Smoke modelling for the atrium under multiple fire scenarios
- Evacuation time modelling under peak load
- Enhanced smoke clearance and early alarm activation

Building Control reviews the evidence and:

- Requests a peer review by an independent fire consultant
- Requires additional compartmentation or suppression systems
- Conditions approval on as-built validation of airflow rates and damper operation

Their scrutiny ensures that the life safety risk of the design is not simply accepted on good faith—but validated through performance logic, tested assumptions, and enforceable controls.

Balancing Innovation with Responsibility

In the PBBD paradigm, the success of a building hinges on its ability to perform—not just functionally, but safely and reliably. Building Control professionals are the guardians of this performance mandate. They do not block innovation; they refine it, validate it, and anchor it in public trust.

By insisting on rigorous, proportionate, and transparent assessment of life safety and structural integrity, Building Control ensures that performance-based design never trades creativity for compliance, or progress for protection. Their role is to ensure that the freedom to design is matched by the responsibility to perform, safely, legally, and ethically.

Facilitating Compliance Through Dialogue and Guidance: Building Control as a Strategic Partner in Design Innovation

In traditional regulatory frameworks, Building Control has often been viewed as a reactive authority, engaging only at key checkpoints to assess whether a project complies with prescriptive standards. However, in the context of Performance-Based Building Design (PBBD), this role becomes much more proactive and collaborative. Because PBBD allows for alternative means of compliance, there is no singular path to approval; instead, compliance becomes a negotiated process informed by evidence, expert judgment, and contextual risk.

In this evolving environment, Building Control serves not only as an enforcer of regulations, but as a facilitator of intelligent, performance-focused solutions. Their advisory role is vital to ensuring that innovative or unconventional strategies are introduced responsibly, efficiently, and with full regulatory transparency.

Navigating a Non-Linear Path to Compliance

Performance-based projects do not follow a fixed script. They often involve:

- Emerging technologies not yet addressed in existing standards
- Complex occupancy profiles or hybrid use types
- Site-specific challenges, such as historic fabric, constrained access, or sustainability requirements
- Ambitious performance targets, such as net-zero energy, passive house standards, or modular construction

These factors mean that the route to compliance must often be customised, with intermediate milestones, ongoing engagement, and early intervention to avoid late-stage disputes. Building Control, when engaged early, can help guide the project team toward viable and defendable solutions before formal submission.

Building Control's Facilitative Functions in PBBD

Providing Early-Stage Guidance on Alternative Compliance Pathways

One of the most valuable contributions Building Control can make is during RIBA Stages 2–3, when design options are being developed. By discussing the acceptability of performance-based strategies before full design commitment, they help teams:

- Identify which deviations from guidance are likely to require enhanced justification
- Suggest acceptable forms of supporting evidence (e.g., modelling tools, test standards, certifications)
- Recommend incremental validation (e.g., phased peer reviews or prototype testing)

This pre-emptive approach can significantly reduce design risk and minimise the likelihood of major revisions later in the process.

Hosting Design Compliance Workshops or Pre-Submission Consultations

Progressive Building Control bodies increasingly offer compliance clinics, workshops, or early-stage consultation sessions. These sessions bring together:

- Architects and lead designers
- Fire and structural engineers
- Sustainability consultants
- MEP designers and BMS specialists

Topics typically include:

- Interpretation of regulatory intent for specific building types
- Discussion of performance-based strategies and potential red flags
- Review of precedent projects or test data
- Coordination of compliance documentation strategy

This multi-disciplinary forum encourages open, solution-focused dialogue, enabling smoother submissions and shared ownership of compliance.

Advising on Documentation, Testing, and Evidence Standards

Performance-based strategies often rely on large volumes of data, modelling outputs, test certificates, simulation screenshots, technical datasheets, etc. Building Control provides clarity on:

- What level of evidence is required (e.g., full CFD model vs simplified zone model)
- Which standards are acceptable (e.g., ISO, BS EN, UKAS-accredited tests)
- How information should be structured, presented, and validated
- How long evidence remains valid and what change triggers revalidation

This guidance helps prevent information overload or incomplete submissions, and supports teams in preparing robust, professional, and concise approval packages.

Offering Contextual Insight from Previous Projects

Experienced Building Control professionals draw on a wealth of historical knowledge. They can:

- Reference how similar strategies were previously approved (or rejected)
- Highlight lessons learned from past failures or enforcement cases
- Suggest conservative mitigations that have enabled borderline strategies to gain approval

This institutional insight supports consistency, fairness, and transparency, especially when dealing with novel approaches or materials.

Benefits of Early and Active Building Control Engagement

By facilitating compliance through ongoing guidance, Building Control helps:

- Streamline the approval process, reducing review timelines and rework
- Improve design quality, by encouraging technically sound alternatives

- Promote legal defensibility, through well-documented performance justifications
- Support team confidence, allowing designers to innovate within clear boundaries
- Reinforce a culture of shared accountability, where regulation is not a barrier but a partner in better building delivery

These dynamic fosters a more intelligent, less adversarial compliance culture, one that rewards clarity, proactivity, and cross-disciplinary collaboration.

Enabling a Culture of Responsible Innovation

In the PBBD context, regulatory frameworks must evolve in tandem with the buildings they seek to govern. Materials, systems, and user expectations are changing rapidly, and Building Control professionals, when acting as facilitators, play a key role in enabling innovation without sacrificing standards.

For example:

- A design team exploring mass timber in a high-rise office can work with Building Control to phase in fire testing, compartmentation design, and smoke control modelling, ensuring that approvals are built on progressive evidence.
- A team pursuing zero-carbon targets using phase change materials may require feedback on thermal modelling, condensation risks, and lifecycle replacement strategies, all of which Building Control can advise on without delaying progress.

Such engagement ensures that emerging technologies are trialled responsibly, documented rigorously, and applied in a manner that maintains regulatory confidence and public safety.

Dialogue as a Driver of Regulatory Clarity and Design Confidence

The regulatory process in PBBD is no longer a static barrier at the end of the design timeline. It is an ongoing, constructive dialogue, a partnership between innovation and oversight, creativity and caution, ambition and accountability.

Building Control, through its facilitative and advisory role, helps project teams move through this dialogue with confidence. By providing early feedback, structured consultation, and technical clarity, they ensure that performance-driven designs do not face avoidable roadblocks, and that emerging practices can flourish within a legally defensible framework.

In this way, Building Control becomes not just an authority, but a collaborator in the delivery of safe, sustainable, and high-performing buildings.

Supporting Whole-Life Compliance and Post-Completion Certification: Securing Performance Beyond Design

The delivery of a building is not complete when construction ends, it is complete when it performs in alignment with its regulatory, safety, and user objectives, and when it receives the legal endorsement to operate. In the context of Performance-Based Building Design (PBBD), where success is tied to functional outcomes rather than rigid compliance templates, the role of Building Control does not stop at the drawing review stage. Instead, it extends well into the construction, commissioning, and post-completion phases, ensuring that what was approved in principle is faithfully realised in practice.

This continued involvement is not merely procedural. Building Control acts as a custodian of whole-life compliance, overseeing the final alignment between design intent, as-built conditions, and operational readiness. Their scrutiny ensures that buildings do not just meet the letter of the law, they meet its intent, maintaining safety, usability, and performance integrity throughout their lifecycle.

Construction Oversight: Verifying Execution Against Approved Design

During the construction phase, Building Control officers monitor whether the build aligns with the approved documentation and compliance strategies. This involves:

- On-site inspections to verify key stages of work, particularly for fire safety features (e.g., compartment walls, cavity barriers, penetrations)
- Confirmation of material specifications, particularly where non-standard products or systems have been used under performance-based justifications
- Review of construction methods, ensuring that workmanship standards support the functional outcomes of the approved strategy

In PBBD, the gap between design and delivery must be tightly managed, as performance relies on systems being installed and integrated exactly as intended. For example, smoke ventilation systems that were modelled for air changes and extraction rates must be installed, commissioned, and tested accordingly—not substituted or downgraded without reassessment.

Commissioning and As-Built Validation: Proving Performance

A core principle of PBBD is that performance must be demonstrated, not assumed. As such, Building Control plays a key role in reviewing the evidence that supports post-installation validation. This includes:

- Commissioning certificates and test results for active systems (e.g., fire alarms, pressurisation systems, BMS-controlled lighting)
- As-built energy modelling, updated to reflect final configurations and product selections
- Post-construction air permeability (blower door) test reports, which confirm the integrity of the building envelope against energy and ventilation targets
- Acoustic test reports, verifying sound insulation levels for separating walls and floors in residential or educational buildings
- Access audits, to ensure inclusive design principles have been maintained from planning to final fit-out

In PBBD projects, where alternative solutions may have been accepted in lieu of standard specifications, this as-built evidence is essential to validate that the alternative approach performs as proposed.

Final Sign-Off: Legal Certification and Operational Readiness

Only when Building Control is satisfied that the building:

- Complies with the functional requirements of the Building Regulations
- Aligns with the approved drawings, strategies, and technical justifications
- Demonstrates performance through robust and verifiable documentation do they issue the Final Certificate or Completion Certificate.

This certificate:

- Legalises occupation or use of the building
- Enables insurance coverage and property registration
- Serves as a regulatory record that the building has been delivered in accordance with national legislation

In some cases, particularly complex buildings or phased handovers, this may be supported by:

- Partial Completion Certificates (e.g., per building zone or system)
- Sectional approvals tied to staged occupation strategies
- Performance warranties or declarations that carry through into operational management

Supporting Futureproofing and Lifecycle Accountability

Building Control's oversight at completion does not simply serve the present, it helps to future-proof the building by ensuring that all compliance and performance data is:

- Accurately recorded and traceable
- Integrated into digital handover platforms or asset management systems
- Aligned with regulatory expectations for future refurbishment, adaptation, or compliance audits

For example, in net-zero buildings or projects seeking sustainability certification (e.g., BREEAM, LEED, NABERS), final performance data reviewed by Building Control may also contribute to:

- Energy certification processes (e.g., EPC, DEC)
- ESG reporting and corporate responsibility
- Asset valuation and lifecycle planning

Thus, Building Control's sign-off is not the end of a process, it is the foundation of a building's regulatory identity for decades to come.

Conclusion: Building Control as the Final Gatekeeper of Performance Assurance

In performance-based design, the gap between aspiration and reality is bridged not only by good design and competent construction, but by rigorous, structured, and independent validation. Building Control officers provide that validation, through inspections, evidence review, and post-completion certification.

Their role at project close-out ensures that buildings do not simply get built, but get built to perform, legally, functionally, and sustainably. This final step, backed by credible documentation and transparent processes is what transforms design ambition into operational assurance, and what allows building stakeholders to occupy, operate, and insure their assets with full confidence.

Building Control as a Strategic Enabler of Performance-Based Design

In the evolving discipline of Performance-Based Building Design (PBBD), the traditional roles of project stakeholders are being redefined. Architects are now system integrators, engineers are scenario modellers, and clients are long-term performance investors. Within this dynamic environment, Building Control professionals are emerging not merely as gatekeepers of compliance, but as critical strategic enablers, facilitators of innovation, interpreters of regulation, and guardians of public safety.

Where once the regulatory process was viewed as an obstacle to creativity, PBBD demands that regulation be part of the creative process itself. Building Control officers enable this by applying functional judgment rather than enforcing rigid prescriptions. They assess not just what a design does, but how it performs in fire, in structure, in use, and across its lifecycle. Their knowledge ensures that emerging materials, new methods of construction, and unconventional design approaches can be explored without compromising the foundational principles of safety, access, and resilience.

Balancing Flexibility with Accountability

Performance-based design requires a balance between freedom of solution and rigour of validation. Building Control professionals provide this balance by:

- Interpreting the intent behind regulations, ensuring that alternative methods achieve equivalent (or superior) outcomes.
- Assessing technical evidence such as simulations, risk assessments, and third-party certifications.
- Identifying proportional responses to risk, aligning regulatory expectations with the complexity, use, and scale of the building.

This approach promotes contextual flexibility while reinforcing accountability, a balance that is vital in an era where design ambitions often extend beyond the boundaries of traditional guidance.

Collaboration Over Confrontation

A key strength of Building Control in PBBD lies in their ability to collaborate across disciplines. From early-stage compliance workshops to final performance verification, they:

- Engage with designers, engineers, and contractors to shape viable alternative strategies.
- Offer early insights that reduce the risk of late-stage rework.
- Provide continuity between regulatory approval and on-site execution.

In this way, Building Control contributes to a culture of shared responsibility, where regulatory compliance is no longer a reactive milestone but a design driver, a proactive process that enhances clarity, coordination, and confidence throughout the project lifecycle.

Securing Performance for the Long Term

Building Control's role extends beyond the immediate scope of design sign-off. Their engagement through construction, commissioning, and certification ensures that the built outcome matches the approved vision, and that all performance assumptions are validated, documented, and legally supported.

This final act of regulatory oversight, supported by inspections, test reviews, and compliance documentation that secures the building's operational legitimacy. It enables:

- Safe occupancy and use
- Insurance coverage and asset valuation
- Lifecycle maintenance planning
- Legal defensibility in the event of future disputes or regulatory review

Through their consistent engagement, Building Control officers safeguard the long-term performance integrity of the built environment.

Enabling Innovation Through Trust

Perhaps most importantly, Building Control professionals provide a bridge between innovation and public trust. Their ability to understand emerging technologies, assess new risk profiles, and approve alternative design paths ensures that building innovation can proceed responsibly, with confidence in its safety and legal soundness.

Whether approving fire-engineered solutions, evaluating offsite modular construction, or validating ultra-low carbon strategies, Building Control makes it possible for performance-led design to advance the potential of architecture without undermining its social and regulatory obligations.

A Foundation for the Future of Design

As PBBD continues to expand across sectors, driven by climate targets, digital tools, and user-centric design philosophies, the role of Building Control will only become more critical. Their technical scrutiny, regulatory expertise, and collaborative mindset will underpin the delivery of:

- Net-zero buildings
- Smart, sensor-driven environments
- Inclusive, resilient, and adaptive urban infrastructure

In this landscape, Building Control is not a passive reviewer, but an active contributor to the quality, safety, and performance of the built environment.

Final Reflection: Beyond Compliance, Toward Contribution

In summary, Building Control in the PBBD paradigm is not just a guardian of rules, but a co-author of outcomes. By ensuring that performance-based solutions are compliant, achievable, and well-documented, they transform regulatory oversight into a form of stewardship, one that champions both public interest and design excellence.

Their role is essential to making PBBD both legitimate and practical: safeguarding life, enabling innovation, and reinforcing the credibility of performance as a measurable, meaningful design philosophy. Through their guidance, dialogue, and expertise, Building Control professionals ensure that buildings of the future are not just imagined boldly, but delivered responsibly.

Building Control as a Strategic Enabler of Performance-Based Design

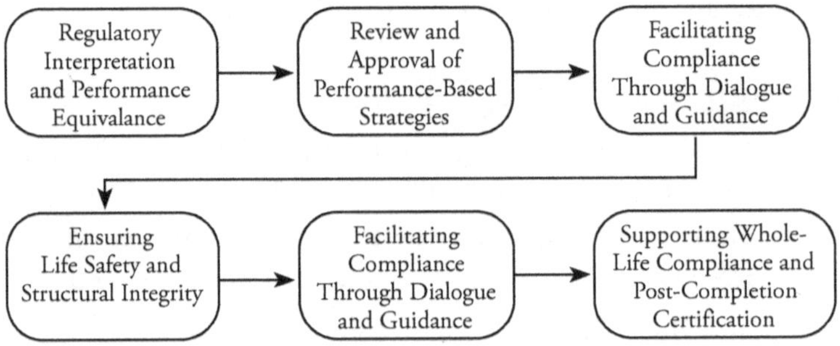

Chapter 12

Role of the Fire Authority in Performance-Based Building Design

Introduction: Fire Authority Involvement in Performance-Based Building Design

In the rapidly advancing field of Performance-Based Building Design (PBBD), regulatory compliance is increasingly judged not by rigid adherence to prescribed solutions, but by the ability of the building to achieve defined, measurable functional outcomes. These outcomes may relate to structural integrity, energy performance, user comfort, or, critically, life safety in the event of fire. Among all the domains within PBBD, fire safety remains one of the most complex, scrutinised, and risk-sensitive, due to its direct implications for occupant survival, firefighter protection, and emergency response efficacy.

Traditional code-based approaches to fire safety, while reliable and proven, often limit design flexibility, especially in innovative or non-standard buildings. By contrast, PBBD enables designers and fire engineers to propose engineered fire strategies that may include extended travel distances, single escape stairs, bespoke smoke control systems, or the substitution of fire-resisting materials, so long as these alternatives are justified through robust evidence and demonstrate compliance with the functional requirements of national building regulations (e.g. Part B of the Building Regulations in England and Wales).

However, this flexibility demands heightened oversight, independent review, and operational realism. It is within this context that the Fire and Rescue Authority (often referred to simply as the Fire Authority) assumes a critical statutory and advisory role. Design teams may develop advanced simulation models and propose innovative solutions, but the Fire Authority must ensure that these strategies can be effectively implemented during a real fire event, under the pressures and constraints of a live emergency.

A Statutory Role Grounded in Regulation and Practice

The Fire Authority is formally designated as a statutory consultee under Regulation 12 of the Building Regulations 2010 (as amended). This means that, while they do not possess the power to approve or reject an application outright, they are legally entitled to review and comment on fire safety matters during the Building Control approval process. They also enforce the Regulatory Reform (Fire Safety) Order 2005. Their input becomes especially important when:

- Alternative fire strategies are used in place of prescriptive methods
- The development has unusual risk factors, such as high-rise configurations, atriums, mixed-use occupancy, or large assembly areas
- Firefighting operations may be compromised by limited access, complex geometry, or unconventional systems

In these cases, the Fire Authority offers specialist operational insight that complements the technical assessments made by Building Control and the fire engineer. Their feedback may prompt design revisions, further testing, or additional safety provisions. Equally, it can serve to validate and support a well-justified performance-based solution, giving confidence to both the project team and the approving authority.

Focusing on Life Safety and Firefighting Operability

Unlike purely regulatory bodies, the Fire Authority brings a unique perspective to fire strategy evaluation, one rooted in real-world emergency response experience. Their core concerns extend beyond theoretical compliance and include:

- Whether the proposed escape strategy allows occupants to safely evacuate before conditions become untenable
- Whether firefighters can access the building quickly and safely, with clear lines of travel and effective firefighting infrastructure
- Whether emergency communication and control systems are intuitive, accessible, and functionally integrated
- Whether new technologies or layouts will complicate fire suppression, search and rescue, or evacuation coordination

This perspective is invaluable in PBBD, where engineered fire strategies must not only perform well in simulations but also translate into operationally viable solutions during the high-stress, time-critical reality of a fire event.

A Partner in Proactive Design, Not Just Passive Review

As the boundaries of architectural design continue to expand, through greater building heights, more ambitious sustainability goals, and increased reliance on digital systems, the Fire Authority's role is evolving as well. Increasingly, fire safety planning is viewed as a collaborative process, with the Fire Authority engaging in:

- Early-stage design consultations
- Pre-application strategy reviews
- Joint planning sessions with Building Control and fire engineers

This early and ongoing involvement supports the iterative refinement of fire strategies, helping to avoid late-stage objections and ensure that buildings are not just compliant but resilient and operable in the face of emergency scenarios.

Bridging Design and Response

In summary, the Fire Authority is an essential partner in performance-based building design, not as an obstacle to innovation, but as a safeguard against complacency. Their role ensures that fire strategies are not only designed for theoretical performance, but for real-world execution under the most demanding conditions.

By reviewing fire strategies through the lens of emergency planning, occupant protection, and operational feasibility, the Fire Authority acts as a bridge between design ambition and life safety assurance. Their contributions reinforce the core objective of PBBD: to enable buildings that are not only high-performing and visually compelling, but safe, responsive, and prepared for the unexpected.

Core Functions of the Fire Authority in PBBD

Review of the Fire Safety Strategy: Ensuring Operability, Life Safety, and Practicality in Performance-Based Approaches

In performance-based building design (PBBD), the development of a Fire Safety Strategy (FSS) is a foundational element—serving not only as a technical document that demonstrates compliance with regulatory requirements, but also as a blueprint for occupant protection, property resilience, and emergency response. Unlike prescriptive design methods that dictate fixed parameters such as maximum travel distances or stair configurations, PBBD empowers designers and fire engineers to develop tailored strategies that respond to the unique characteristics of the building and its occupants.

It is within this context that the Fire Authority's review of the Fire Safety Strategy becomes indispensable. As a statutory consultee, the Fire Authority examines the FSS through a practical, operational lens, assessing not only whether the proposed strategy complies in principle, but whether it can be effectively executed in the real-world conditions of a fire event.

This review process is comprehensive and typically focuses on the following core elements:

Means of Escape for Occupants

A primary concern for the Fire Authority is ensuring that all building users can safely and efficiently evacuate in the event of fire. This involves detailed scrutiny of:

- **Egress routes and travel distances:** Are these within tolerable limits based on the building's risk profile and occupancy type? If they exceed standard guidance, are appropriate compensatory measures (e.g. enhanced detection or pressurisation) in place?
- **Stair and exit provisions:** Are sufficient escape stairs and final exits provided? How do evacuation routes support both phased and simultaneous evacuation models?
- **Protected escape paths:** Are escape routes enclosed and protected from fire and smoke ingress? Have the required fire resistance ratings for doors, walls, and lobby enclosures been provided?
- In PBBD, where longer travel distances or fewer escape routes may be proposed to enhance spatial efficiency or architectural intent, the Fire Authority assesses:
    - Whether egress models (e.g. BuildingExodus, Pathfinder, MassMotion) demonstrate tenable conditions for occupants throughout the evacuation timeline.
    - Whether fire growth rates, occupant densities, and behavioural assumptions are realistic and aligned with the building's use case.

Compartmentation and Fire Resistance

The Fire Authority ensures that the strategy provides adequate fire containment, limiting the spread of heat, smoke, and flame both vertically and horizontally. This includes:

- Assessment of fire-resisting construction (e.g. walls, floors, penetrations) and fire-stopping details
- Compartmentation strategy for different uses within the same building (e.g. retail below residential)
- Limitation of unprotected voids and service penetrations, especially in high-rise and complex buildings

In performance-led designs, such as open-plan offices, interconnected atria, or tall residential towers, where compartmentation may be strategically relaxed, the Fire Authority evaluates:

- The effectiveness of alternative containment methods (e.g. smoke curtains, suppression systems, active controls)
- The integrity of compensatory measures in maintaining boundary conditions for safe evacuation and firefighter operations

Detection, Alarm, and Suppression Systems

The Fire Authority pays close attention to the nature, placement, and integration of:

- Automatic fire detection systems (e.g. addressable point detectors, beam detectors, aspirating smoke detectors)
- Alarm notification systems (e.g. phased alarms, voice evacuation systems)
- Active suppression systems (e.g. sprinklers, water mist, gaseous suppression)
- Fire control interfaces (e.g. fire alarm panels, graphical user interfaces, BMS integration)

In PBBD, these systems often play a critical role in offsetting design deviations, such as:

- Reduced stair cores (enabled by early warning and fast suppression)
- Extended escape distances (supported by detection-triggered smoke ventilation)

The Fire Authority assesses whether:

- System specifications are appropriate to the building's fire load, geometry, and risk category
- Detection coverage ensures early fire localisation and system activation
- Alarm systems are comprehensible, accessible, and free from nuisance activation risks
- System redundancy, power backup, and maintenance protocols are clearly established

Access for Firefighting Operations

Effective fire service response is contingent upon appropriate firefighter access, which is a priority in every Fire Authority review. Key areas of focus include:

- **Appliance access routes:** Do emergency vehicles have unimpeded access to the building perimeter, and is sufficient hardstanding provided for aerial operations or pumping?
- **Internal access provisions:** Are dry risers, wet risers, firefighting shafts, and stair cores correctly positioned and accessible within recommended distances?
- **Evacuation lifts and fire lifts:** Where provided, are they protected, robust, and fully supported by the building's power and alarm infrastructure?
- **Wayfinding and communication systems:** Are building plans available in fire control rooms? Are zone maps, signage, and smoke control panels intuitive and compatible with standard fire service procedures?

In buildings with complex internal layouts or automated fire protection systems, the Fire Authority seeks assurance that their crews will not face operational delays, spatial confusion, or system incompatibility during critical response windows.

Evaluating Engineered Alternatives and Compensatory Measures

Where performance-based strategies propose non-standard fire safety arrangements, the Fire Authority must evaluate whether the intended outcome, safe evacuation and effective firefighting remain achievable. This may involve reviewing:

- Smoke modelling outputs (e.g. SmartFire, FDS simulations)
- Fire growth rate assumptions under various ignition scenarios
- Occupant behaviour assumptions for different user groups (e.g. staff, visitors, vulnerable users)
- Redundancy and fail-safes in automated systems

Critical questions include:

- Are intervention timelines for the fire service realistically considered in the modelling?
- Are smoke and temperature conditions tenable for egress and fire attack along the intended access routes?
- Do the compensatory measures genuinely offer equivalent or better protection than the prescriptive baseline?

If there is insufficient clarity, the Fire Authority may:

- Recommend further modelling or fire testing
- Request peer review of the fire strategy
- Suggest enhancements such as added suppression coverage, secondary access points, or more robust evacuation signage

Impact of Fire Authority Feedback

While the Fire Authority does not formally approve the design, their feedback carries significant influence and may:

- Prompt substantive design revisions
- Lead to the inclusion of enhanced mitigation measures
- Affect the conditions of Building Control approval
- Shape the operational strategies of the fire and rescue service post-completion

Engaging with the Fire Authority early in the design process, particularly during the fire strategy development and pre-application phases can help mitigate risks, resolve conflicts proactively, and streamline the compliance journey.

Firefighter Access and Facilities: Ensuring Operational Readiness and Tactical Viability

In the event of a fire, the speed, safety, and effectiveness of the fire and rescue service's response are heavily influenced by the quality of firefighter access and supporting infrastructure designed into the building. For this reason, Firefighter Access and Facilities form a cornerstone of the Fire Authority's review in Performance-Based Building Design (PBBD) projects.

Unlike code-driven designs where access parameters are clearly prescribed and often conservative, PBBD enables greater flexibility in site layout, form, and construction logic. However, this flexibility must not compromise the tactical viability of firefighting operations. Fire crews need to arrive, set up, enter, navigate, and suppress fire quickly, often under intense heat, smoke, and structural degradation. The Fire Authority ensures that, regardless of the design approach, critical firefighter access principles are respected and practically achievable.

Core Areas of Assessment for Firefighter Access

Vehicle Access and Appliance Positioning

The Fire Authority reviews whether:

- Fire appliances can get within 45 metres of any part of the building's footprint, as per Approved Document B (ADB) requirements, or if PBBD alternatives provide equal or better coverage.
- Access roads are suitable for the weight, size, and turning capabilities of modern fire appliances, including aerial ladder platforms and rescue units.

- Hardstanding areas are level, load-bearing, and free from obstructions (e.g. street furniture, bollards, soft landscaping).
- Turning heads and tracking radii enable safe manoeuvring for fire trucks, particularly at dead-ends, under crofts, or podium decks.

In PBBD projects with constrained urban plots, landscaped perimeters, or complex site topography, the Fire Authority scrutinises swept path analyses, ground gradient tolerances, and appliance staging areas to verify that real-world response timelines are not impaired.

Dry and Wet Risers, Fire Mains, and Hydrant Networks

For internal firefighting, the building must support safe and efficient water delivery to the upper or lower floors. The Fire Authority assesses:

- **Dry risers (for buildings 18m–60m in height):** Are they provided with inlet points at accessible and unobstructed locations? Are they clearly marked, with direct connection to firefighting lobbies?
- **Wet risers (required above 60m):** Is the pump infrastructure reliable, dual-powered, and tested under operating pressures?
- **Fire mains and hydrant locations:** Do they provide adequate coverage for the full building perimeter and interface effectively with external suppression efforts?
- **Compatibility of connections:** Are fire hydrant outlets, valve types, and riser fittings compatible with local fire and rescue equipment?

Any deviation from standard configurations, such as reduced riser provision or reliance on portable pumps, must be carefully justified, with flow rate modelling, operational time assumptions, and risk mitigation strategies.

Firefighting Shafts and Protected Access Routes

Tall and complex buildings must incorporate firefighting shafts, providing:

- A fire-rated stairwell
- A firefighting lift with dual electrical supply
- A firefighting lobby with smoke protection

The Fire Authority ensures that these shafts:

- Are positioned to optimise travel distances for hose lines (typically no more than 60m from all points on the floor plate)
- Are accessed directly from fire appliance parking zones, without reliance on internal circulation routes or atria

- Remain tenable for crew ingress, with pressurisation, ventilation, or smoke control systems maintaining visibility and air quality

In buildings with alternative layouts or phased evacuation zones, the Fire Authority will often request tenability modelling and access route visualisation to confirm that operational assumptions remain valid under fire conditions.

Evacuation Lifts and Refuge Areas

In modern PBBD projects, especially those seeking inclusive design or high sustainability ratings, the use of evacuation lifts is increasingly common. The Fire Authority reviews:

- Whether lifts intended for evacuation or fire service use are fire-rated, water-resistant, and connected to reliable backup power
- Whether refuge areas adjacent to lifts are:
    - Sufficiently sized for wheelchair users and carers
    - Provided with two-way communication systems
    - Adequately protected against fire and smoke

Where fire-fighting lifts double as evacuation lifts, the design must include:

- Appropriate fire detection interlocks
- Dedicated lift control overrides for fire crews
- Robust system testing and commissioning protocols

The Fire Authority ensures that lift usage does not introduce new operational vulnerabilities, particularly in high-rise buildings, care facilities, or complex mixed-use developments.

Deviations and Alternative Approaches in PBBD

Performance-based approaches may propose deviations from conventional access arrangements. These might include:

- Reliance on aerial appliances for upper-storey access instead of internal stairs
- Use of mobile pumping units to compensate for reduced riser infrastructure
- Reconfigured internal layouts with horizontal escape corridors rather than vertical shafts

In these cases, the Fire Authority evaluates:

- The operational feasibility of the alternative method under stress, smoke, and time-critical conditions
- The impact on firefighter safety, access time, and hose reach

- The reliability and maintainability of any technology or system relied upon

This evaluation often requires additional evidence, such as:

- Operational modelling or real-time fire drills
- Scenario walkthroughs with fire crews
- Peer-reviewed design studies

If necessary, the Fire Authority may request that certain features be reinstated or augmented to preserve operational functionality.

Case Example: Access Strategy in a Constrained Urban High-Rise

A high-rise residential tower in a tight city centre location proposed:

- One firefighting shaft located in the building core
- Vehicle access to only two sides of the site
- Suppression coverage via sprinklers and wet risers

The Fire Authority raised concerns about:

- Excessive internal travel distances from the riser outlets
- Limited ground-level appliance positioning
- Inadequate evacuation lift resilience in case of power failure

In response, the design team:

- Reconfigured the riser layout to reduce hose run distances
- Strengthened the fire lift lobby with independent power and pressurisation
- Agreed to provide a firefighter staging area with remote comms interface

This collaborative solution enabled the performance-based layout to proceed without compromising firefighter access or safety.

Designing for Response, Not Just Escape

In PBBD, designing for safe occupant evacuation is only half the equation. A truly resilient building must also support the rapid, informed, and safe response of emergency services, even when that building challenges conventional forms.

The Fire Authority's review of firefighter access and facilities ensures that the building can be effectively engaged by crews under adverse conditions. Whether it's delivering hose water to the 20th floor, accessing a basement car park under heavy smoke, or managing phased evacuation in a smart building, their operational insight is essential to bridging design intent with emergency realism.

By ensuring these systems are not only compliant on paper but tactically viable on the ground, the Fire Authority plays a pivotal role in shaping buildings that protect life, during both evacuation and intervention.

Emergency Planning and Operational Preparedness

The Fire Authority is not limited to design review. It also has a role in emergency preparedness, especially for:

- Large-scale developments (e.g., airports, hospitals, stadiums)
- Complex buildings with phased evacuation, zoned fire suppression, or occupant management systems
- Sites with unique fire risks, such as hazardous materials or energy storage systems

Their input helps ensure that emergency response plans are integrated into the building's design, supporting:

- Familiarisation for fire crews
- Pre-incident planning
- On-site signage and information systems (e.g., fire control panels, as-fitted plans)

In PBBD, where systems may operate differently from standard expectations, ensuring that responders understand the building's behaviour under fire conditions is vital for safe intervention.

Consultation and Feedback Mechanisms: Enhancing Performance Strategy Through Engagement and Expertise

In the regulatory landscape of Performance-Based Building Design (PBBD), the delivery of an effective and compliant fire strategy hinges not only on engineering analysis but also on early, meaningful, and iterative consultation with key stakeholders, foremost among them, the Fire Authority. Although the Fire Authority does not possess the statutory power to approve or reject a project under building regulations, their formal role as a statutory consultee under Regulation 12 of the Building Regulations 2010 (as amended) grants their input significant weight in the approval process.

This position enables the Fire Authority to influence design outcomes, highlight operational concerns, and shape the fire safety narrative through a series of structured feedback channels. Their engagement ensures that performance-based fire strategies are not only technically valid but practically executable in the dynamic, high-risk context of real-world emergencies.

Regulatory Weight of Fire Authority Feedback

While Building Control remains the approving authority for compliance with Part B (Fire Safety) of the Building Regulations, Fire Authorities are expected to be consulted where public safety, firefighter access, or fire strategy complexity may impact operational response. Their formal responses carry the following implications:

Objections or Conditions Precedent to Approval

Where significant life safety or operational risks are identified, the Fire Authority may issue:

- Objections to critical aspects of the fire strategy (e.g. inadequate firefighting shafts, overly complex evacuation routes, or high-risk atrium designs)
- Conditional responses, requiring remedial actions such as revised compartmentation layouts, additional suppression systems, or redesigned access arrangements

These comments must typically be resolved to the satisfaction of Building Control before the project can receive final sign-off.

Recommendations for Operational Enhancements

Even when a design meets minimum compliance thresholds, the Fire Authority may recommend:

- Enhanced wayfinding for responders
- Additional refuge areas and/or communication systems
- Upgraded fire lift configurations for mixed-use or tall buildings
- Installation of override panels, real-time zoning displays, or fire control rooms

These recommendations, while not legally binding, often guide developers towards best practice implementation, especially in buildings with high public occupancy or critical infrastructure functions.

Requests for Further Information

To validate the fire strategy and confirm operational feasibility, the Fire Authority may request:

- Compartmentation layouts and fire-resistance ratings
- Time-temperature curves and tenability analyses
- Egress simulations, especially for non-standard geometries or high-density occupancies

- Fire growth rate assumptions and detection response timelines
- Details of fire alarm logic, phased evacuation protocols, or control system integrations

Such requests reflect the Fire Authority's need to understand the building holistically, not just from a regulatory perspective, but from an emergency response standpoint.

Modes of Engagement and Collaboration

In PBBD, the Fire Authority's role is increasingly moving away from reactive review and towards proactive, co-creative collaboration with design teams. This is especially important in complex or innovative developments where assumptions, systems, or sequences may deviate significantly from conventional designs.

Pre-Application Meetings

These early-stage consultations enable the Fire Authority to:

- Understand the design concept and proposed fire strategy direction
- Flag foreseeable concerns around access, evacuation, or intervention
- Clarify evidence requirements, modelling standards, or peer review expectations
- Offer guidance on suitable control interfaces or digital system integration

By engaging early, the design team avoids investing time and resources into approaches likely to face rejection or major revision later.

Design Team Workshops

For large or complex projects, Fire Authorities often attend multi-disciplinary design reviews, where architects, engineers, surveyors, and project managers collaborate to align performance goals. During these workshops, the Fire Authority can:

- Provide live feedback on layout configurations, control room locations, or fire main routing
- Coordinate assumptions across fire modelling, structural strategy, and mechanical services
- Ensure clarity around evacuation responsibilities, communications, and interface protocols

This collaborative format supports shared understanding and cross-disciplinary accountability.

Joint Fire Safety Audits

In certain high-risk or high-profile projects (e.g., hospitals, transport hubs, high-rise residential schemes), Fire Authorities may conduct joint audits with Building Control during design development or post-commissioning. These audits allow for:

- Physical review of access routes, fire-fighting shafts, and fire alarm panels
- Validation of as-built compliance with the Fire Safety Strategy
- Dialogue around ongoing maintenance, crew familiarisation, and fire safety management plans

These joint reviews promote whole-building resilience and provide an opportunity to fine-tune systems before occupancy.

Strategic Value of the Consultation Process

The consultation and feedback mechanisms led by the Fire Authority offer numerous advantages for PBBD projects:

- De-risking of approval pathways by pre-emptively resolving fire safety objections
- Increased robustness and credibility of fire strategies through informed critique
- Faster regulatory approvals by streamlining inter-agency coordination
- Enhanced public trust and operational assurance, particularly for tall buildings or high-occupancy facilities

Moreover, regular consultation helps project teams build institutional knowledge about fire service expectations, evolving risks, and innovations that may become mainstream in future design guidance.

Consultation as a Pillar of Fire Safety Assurance

In performance-based design, consultation with the Fire Authority is not an administrative formality, it is a critical phase of design validation and operational alignment. By contributing their real-world experience, response expertise, and statutory perspective, Fire Authorities help shape fire strategies that are not only compliant in theory but deliverable in the heat of an emergency.

Through structured feedback, early engagement, and post-design collaboration, the Fire Authority ensures that performance-based solutions uphold the core tenets of life safety, firefighter protection, and public assurance, making consultation a foundational pillar in the responsible application of PBBD.

Importance of Engagement in PBBD

In PBBD, fire strategies often rely on complex performance modelling—such as:

- Computational Fluid Dynamics (CFD) simulations
- Probabilistic risk assessments
- Occupant movement modelling (e.g., BuildingExodus, Pathfinder, MassMotion)
- Fire load assessments and smoke layer calculations

The Fire Authority reviews these with a focus not only on compliance, but also on operability, asking:

- Can crews safely enter and work in these environments?
- Will smoke management systems hold tenable conditions?
- Are firefighting shafts and stair cores correctly located and supported?
- Are fire control interfaces intuitive and appropriately zoned?

By grounding theoretical models in operational reality, the Fire Authority plays an essential role in bridging the gap between design assumption and emergency execution.

Example Scenario: Fire Authority Influence on Design

In a high-rise residential development proposing a single staircase escape design supported by mechanical smoke extraction and enhanced detection, the Fire Authority raised concerns regarding:

- Evacuation capacity under simultaneous egress scenarios
- Potential delays in fire crew access if the staircase became compromised
- The need for additional smoke lobbies and a protected firefighting lift

Following consultation, the design was modified to include:

- A larger pressurised escape stair
- Enhanced lobby protection
- Supplementary real-time smoke detection linked to fire service panels

This ensured both regulatory compliance and confidence in operational effectiveness.

Conclusion: Fire Authorities as Protectors of Life Safety and Operational Integrity

In the realm of Performance-Based Building Design (PBBD), innovation often drives complexity, and complexity increases the margin for risk. Nowhere is this more consequential than in the domain of fire safety, where assumptions about system performance, occupant behaviour, and structural resilience must translate into predictable, reliable protection during a real-world emergency.

While engineering models, performance simulations, and digital visualisations offer powerful tools for validating fire strategies on paper, they cannot fully capture the dynamic, high-pressure conditions of a live fire incident. In such moments, function must override form, clarity must overcome complexity, and systems must support split-second decisions made by emergency responders. It is in this operational crucible that the Fire and Rescue Authority proves its irreplaceable value.

Bridging Theory and Emergency Response

Fire Authorities are not passive observers of design, they are active participants in safeguarding its consequences. As statutory consultees, they serve as a vital checkpoint in the building design process, providing expert insight that transcends regulatory box-ticking. They ensure that fire strategies are:

- Not only code-compliant, but operable and intuitive in the hands of first responders
- Not only innovative, but proven to withstand fire loads, system failures, or delays in evacuation
- Not only engineered for performance, but structured for survivability, of both occupants and firefighters

Their review functions as a reality filter, grounding abstract strategy in lived operational experience.

Reinforcing Accountability Through Collaboration

The Fire Authority's role reinforces a critical principle: fire safety is a shared responsibility. In performance-led design, success cannot rest solely on the shoulders of the fire engineer or Building Control officer. It demands collaborative alignment between:

- Designers, who conceive strategies and spatial configurations
- Regulators, who interpret compliance and equivalency
- Fire Authorities, who plan and act during emergencies

This triad ensures that fire safety is not just an input or output, but a continuous consideration across the project lifecycle, from concept to occupancy, and into long-term management.

Guardians of Public Trust and Long-Term Resilience

Beyond statutory duties, Fire Authorities play a deeper role as guardians of public trust. They help the public, clients, and insurers believe in the safety of the built environment. In the wake of tragedies, regulatory reform, and rising building heights, their role is increasingly viewed not just as essential—but existential.

In PBBD, their presence assures stakeholders that:

- Alternative solutions are grounded in defensible, real-world logic
- Emergency access and evacuation are practical, inclusive, and coordinated
- Fire protection systems will operate as expected, even under exceptional circumstances

Their involvement enhances resilience, credibility, and assurance, particularly in buildings that push the boundaries of material science, digital control, or architectural form.

Supporting Innovation Through Realism

Contrary to perceptions of being risk-averse or conservative, Fire Authorities are often enablers of responsible innovation. Their input can validate the use of:

- Smart suppression systems
- Engineered egress strategies
- Alternative structural fire protection measures
- Digital smoke control and phased evacuation logic

What they demand is not constraint, but rigour, validation, and operational foresight. In doing so, they ensure that innovation in fire design is not only exciting but executable, not just intelligent, but survivable.

Final Reflection: Operational Integrity as a Design Imperative

Ultimately, the Fire Authority safeguards one of the most important yet intangible qualities of any building: its ability to protect life under extreme conditions. Their contribution ensures that fire strategies are not merely creative artefacts or mathematical proofs, they are tools for saving lives and protecting responders, delivered with integrity and supported by practice.

In the context of PBBD, where design freedom must be balanced by functional accountability, the Fire Authority's role is irreplaceable. They are the custodians of operational integrity, the voice of lived emergency experience, and the bridge between safety theory and tactical execution.

Their early and continuous engagement elevates the fire strategy from a document to a living system of protection, one that stands ready not just for approval, but for reality.

THE FIRE AUTHORITY'S ROLE IN PBBD FROM STRATEGY TO SAFETY

REVIEW OF THE FIRE SAFETY STRATEGY

- Assess alternatives to prescriptive fire codes
- Evaluate evacuation and smoke control strategies
- Validate firefighting access and facilities

CHECK FOR OPERATIONAL FEASIBILITY

- Ensure strategies are functional in real fires
- Consider response times and survivability
- Support life safety and firefighting effectiveness

ENGAGE THROUGH DESIGN PROCESS

- Act as statutory consultee with design teams
- Provide feedback, advice and guidance
- Address concerns via consultation and audits

SAFEGUARD OPERATIONAL INTEGRITY

- Bridge innovative solutions and emergency response
- Build accountability and public confidence
- Elevate fire safety beyond compilance

Chapter 13

Project Example: The "FutureCare" Net-Zero Carbon Hospital

Project Type: 10-storey urban hospital

Location: Major UK city

Objective: To deliver a resilient, inclusive, net-zero carbon hospital that supports patient healing, staff wellbeing, and long-term operational efficiency—using a performance-based approach.

Key Stakeholders and Their Roles

Client / Developer

The NHS Trust acting as the client sets ambitious goals:

- Achieve net-zero operational carbon
- Gain BREEAM Outstanding and WELL Platinum certification
- Create a healing environment with biophilic design and high daylight autonomy
- Ensure resilience in power, HVAC, and emergency systems They allocate funding and commission a team experienced in performance-based delivery.

Project Manager

The PM leads project initiation and ensures:

- Performance goals are translated into technical briefs
- Design iterations are aligned with budget and risk management
- Regulatory approvals are coordinated, particularly around Part L (Energy) and Part B (Fire)
- BIM is integrated across disciplines for design and asset management

Architect

Designs an H-shaped floor plan to maximise natural light and cross ventilation in wards. Uses:

- Passive design strategies for thermal regulation
- Shading and façade articulation to reduce solar gain
- Operable windows with sensors and secure night-purge ventilation
- Circulation routes optimised for patient dignity and staff efficiencyThe layout is reviewed iteratively against fire escape models and daylighting simulations.

Fire Engineer

Develops a performance-based fire strategy with:

- Compensated travel distances in wards offset by mechanical smoke control
- Pressurised fire-fighting shafts and smoke lobbies
- Evacuation models (using Pathfinder) showing safe clearance within 4 minutes
- Fire-resistant glazing for visual transparency between wards and escape routes

Structural Engineer

Designs a hybrid steel and CLT frame, reducing embodied carbon. Key features:

- Finite Element Modelling (FEM) for disproportionate collapse and seismic risk
- Connection design for fire and thermal performance
- Integration with services in exposed soffits for post-occupancy flexibility

Mechanical Engineer

Designs an advanced low-temperature HVAC system:

- Chilled beam cooling and underfloor heating
- Heat recovery ventilation (HRV) for operating theatres
- Demand-controlled ventilation integrated with CO_2 sensors
- Backup cooling loops connected to a ground-source heat pump system

Electrical Engineer

Implements a smart power distribution and lighting system:

- Daylight-linked dimming and occupancy sensing
- Emergency lighting with battery backup and wireless testing
- UPS-fed IT zones for medical equipment and diagnostics
- Real-time energy metering and automated peak load shedding

M&E Engineers (Integrated)

Coordinate services into a digital twin, linking:

- HVAC, lighting, and security via the BMS
- Energy dashboards and maintenance alerts
- Real-time performance feedback used in commissioning and early operation

Sustainability Consultant

Leads carbon modelling and green certification:

- Whole-life carbon assessments showing < 300 $kgCO_2e/m^2$
- Oversees BREEAM, WELL, and NHS Net-Zero targets
- Advises on material selection (e.g., low-VOC, recycled content)
- Runs occupant comfort simulations using PHPP and IES VE

Quantity Surveyor / Cost Consultant

Manages capital and lifecycle costs:

- Runs comparative LCC studies of passive vs active cooling strategies
- Evaluates the ROI of PVs, battery storage, and offsite fabrication
- Assesses operational savings over 25-year NHS maintenance periods

Building Surveyor

Monitors quality and compliance during construction:

- Verifies fire-stopping and compartmentation installations
- Reviews cladding systems for thermal bridging
- Ensures construction follows the as-approved digital fire strategy
- Assesses future maintainability of façade and roof plant

Facilities Manager / End User

Engaged from RIBA Stage 2 to guide:

- Service access points and plantroom layout
- Design of nurse stations and equipment bays
- Feedback on lift capacity, wayfinding, and infection control flows
- Tests BMS alert thresholds and maintenance user interfaces before handover

Building Control

Participates in staged reviews:

- Approves deviations from Part B for smoke control strategy
- Validates energy models and final SAP/BRUKL outputs under Part L
- Requires final commissioning data and as-built documentation for Completion Certificate

Fire Authority

Engages in statutory consultation and design review:

- Requires simulation evidence for smoke control tenability
- Recommends relocation of fire control room for appliance access
- Participates in commissioning fire alarm logic and fire-fighting lift operation
- Signs off on operational plan including crew familiarisation

Outcome

The "FutureCare" hospital:

- Achieves BREEAM Outstanding and WELL Platinum
- Performs at 45 $kWh/m^2/yr$ in operation
- Is fully compliant with Building Regulations through performance-led justifications
- Sets a benchmark for PBBD delivery in the public health sector

Closing Chapter

A Future-Ready Paradigm for Building Design

Performance-Based Building Design (PBBD) represents a pivotal shift in how we conceive, plan, and construct the built environment. As the demands placed upon buildings become more complex, the limitations of traditional prescriptive codes are increasingly apparent. PBBD offers an evolved alternative—one that embraces flexibility, innovation, and real-world performance over rigid compliance. Throughout this book, we have explored the depth and breadth of PBBD, illuminating how its application redefines design, engineering, and construction across all building types.

At its core, PBBD redefines success in the built environment, not through compliance with rigid codes but through the achievement of measurable outcomes that prioritize safety, sustainability, resilience, and operational excellence. This method enables professionals to work backwards from clearly defined objectives: How safe should a building be during a fire? How much energy should it consume annually? What level of seismic performance is expected during a once-in-a-century earthquake? These questions form the basis for performance targets, which in turn guide every decision from concept to completion.

This goal-oriented mindset elevates design flexibility and technological integration. From the earliest design stages, PBBD invites stakeholders to explore innovative solutions, guided by performance criteria rather than constrained by prescriptive rules. It enhances our capacity to integrate renewable energy, optimize material use through life cycle assessment, and maintain high levels of indoor environmental quality. In doing so, PBBD supports the broader societal objectives of reducing environmental impact, achieving carbon neutrality, and creating healthier, more adaptive spaces.

One of the most powerful aspects of PBBD is its encouragement of interdisciplinary collaboration. This book has highlighted the unique and indispensable contributions of various disciplines:

- **Client / Developer** sets the vision, objectives, and funding for the project, defining the overall direction and performance ambitions from the outset.
- **Project Managers** are the linchpin of performance-based projects, orchestrating teams, managing risk, and ensuring that performance criteria are embedded in scheduling, cost control, and procurement processes.

- **Architects** shape the spatial, aesthetic, and functional vision of the building while ensuring that layouts, forms, and materials support performance targets such as thermal comfort, daylight access, egress, and energy efficiency.
- **Fire Engineers** use advanced modelling to assess fire and smoke behaviour, simulate evacuation times, and ensure that fire resistance strategies are tailored to building-specific risks. Their expertise is vital for complex and high-risk structures.
- **Structural Engineers** apply risk-based principles and Load and Resistance Factor Design (LRFD) to ensure the building can withstand seismic, wind, and accidental loads, with performance aligned to defined safety and serviceability goals.
- **Mechanical Engineers** design HVAC systems that meet precise targets for thermal comfort, indoor air quality, and energy efficiency.
- **Electrical Engineers** develop power distribution, lighting, and communication systems that align with energy efficiency, user needs, and future adaptability.
- **Mechanical and Electrical (M&E) Engineers** ensure that integrated systems for heating, ventilation, air conditioning, power, and controls work in harmony to meet performance goals across energy use, carbon reduction, and occupant wellbeing.
- **Sustainability Consultants** perform simulations, modelling, and life cycle assessments to align the project with certifications such as BREEAM, LEED, or WELL, focusing on operational and embodied carbon, resource efficiency, and occupant wellbeing.
- **Energy and Environmental Modellers** conduct dynamic simulations to test building performance under varied conditions, providing crucial data for decision-making on energy consumption, emissions, and environmental response.
- **Cost Consultants / Quantity Surveyors** provide critical financial oversight, evaluating the implications of performance-driven choices and ensuring value for money without compromising long-term outcomes.
- **Building Surveyors** play a vital role in inspecting, validating, and ensuring compliance with performance objectives during construction and occupancy. Their knowledge of material behaviour, building pathology, and regulation helps bridge theory and practice.

- **Facilities Managers / End Users** provide operational insights and post-occupancy feedback, ensuring the building delivers on its intended performance goals over its lifespan.
- **Commissioning Agents** verify that all building systems operate as intended, ensuring that performance standards are met before and after handover.
- **Building Control Professionals** play a regulatory and quality assurance role, ensuring that performance-based solutions meet the intent and safety objectives of the Building Regulations. They review non-standard designs and require evidence-based validation of performance outcomes.
- **Fire Authorities** act as statutory consultees, evaluating fire safety strategies to ensure life safety, firefighting access, and emergency procedures are appropriate for the intended use and building type.

This collective endeavour transforms PBBD from a theoretical framework into a practical, high-performing reality. It is a methodology grounded in communication, evidence-based decision making, and a shared commitment to innovation.

The benefits of PBBD are many and far-reaching. It fosters the use of modern materials such as cross-laminated timber, carbon fibre composites, and self-healing concrete. It supports modular and 3D-printed construction techniques that reduce waste and accelerate delivery. It makes room for intelligent building management systems that adapt in real time to changes in occupancy, temperature, and environmental conditions. It is this integration of modern technology with performance-driven intent that positions PBBD as not only relevant, but essential to the future of the built environment.

This future-forward perspective is particularly important in light of global imperatives. The climate emergency demands that buildings operate at net-zero or even net-positive energy levels. Urbanization challenges our ability to design compact, high-performance structures in increasingly dense cities. Ageing infrastructure requires adaptable solutions that can be retrofit and maintained over time. PBBD is uniquely positioned to meet these challenges through its ability to tailor solutions, incorporate resilience measures, and extend the useful life of structures through strategic design.

It is also worth recognizing how PBBD complements and builds upon domain-specific expertise, such as that found in my other book, Principles of Fire Engineering – Science, Safety, and Solutions. That book laid the groundwork for a science-based understanding of fire behaviour, evacuation modelling, fire resistance strategies, and risk-informed design. Within the PBBD context, these concepts find expanded application and are integrated with other performance domains, leading to holistic and optimized building solutions.

As we look ahead, the trajectory of building design is clear. Static compliance models are being replaced by dynamic, performance-based approaches that are smarter, more resilient, and more adaptable. The next generation of designers, engineers, and construction professionals will be called upon not just to meet regulations, but to anticipate change, solve complex problems, and create environments that serve human and ecological well-being.

Let this book serve as both a technical guide and a call to action: to embrace the possibilities of performance-based design, to lead the way in interdisciplinary collaboration, and to build not just for today's standards, but for tomorrow's realities. With PBBD, we are not simply building structures, we are shaping the future of the